Lecture Notes in Mathematics

Edited by A. Dold and B. Eckmann

393

Gordon Wassermann

Stability of Unfoldings

Springer-Verlag
Berlin Heidelberg New York Tokyo

Author

Gordon Wassermann
Abteilung für Mathematik, Universität Bochum
Universitätsstr. 150, 4630 Bochum, Federal Republic of Germany

1st Edition 1974
2nd Printing 1986

Mathematics Subject Classification (1970): 57D45, 58C25

ISBN 3-540-06794-9 Springer-Verlag Berlin Heidelberg New York Tokyo
ISBN 0-387-06794-9 Springer-Verlag New York Heidelberg Berlin Tokyo

Printed in Germany

Printing and binding: Beltz Offsetdruck, Hemsbach/Bergstr.
2146/3140-543210

TABLE OF CONTENTS

§ 1.	Preliminaries	1
§ 2.	Finitely determined germs	35
§ 3.	Universal unfoldings	54
§ 4.	Stable unfoldings	84
§ 5.	The seven elementary catastrophes: a classification theorem	116
Appendix:	Thom's catastrophe theory	153
References		162

INTRODUCTION

The concept of stability plays a major role in the theory of singularities. There are several reasons for the importance of this notion. For one, usually the problem of classifying the objects being studied is extremely difficult; it becomes much simpler if one tries to classify only the stable objects. For another, in many cases (though not in all) the stable objects are generic, that is, they form an open and dense set; so in these cases almost every object is stable and every object is near to a stable one; the non-stable objects are peculiar exceptions. But a third reason for the importance of stability is that the theory of singularities has in recent years, especially through the ideas of R. Thom, acquired important applications to the natural sciences; stability is a natural condition to place upon mathematical models for processes in nature because the conditions under which such processes take place can never be exactly duplicated; therefore what is observed must be invariant under small perturbations and hence stable.

Stability notions have been defined for a variety of objects occurring in the theory of singularities: for mappings, for map-germs, for varieties, for vector fields, for attractors of vector fields and so on. In some cases very little is known about the stable objects; in some cases the stable objects have been completely classified. Other cases lie between these extremes; for example, characterizations of stable proper smooth mappings between manifolds and of stable smooth map-germs have been given by Mather; he has also computed the dimensions in which the stable proper mappings are dense in the set of all proper mappings (see [7], [8], [9]).

For smooth real valued map-germs the theory is rather trivial; such a germ is stable if and only if it has at worst a non-degenerate singularity (or is non-singular).

In this paper the main topic is an investigation of several notions of stability for unfoldings. If η is a germ at the origin of a smooth real-valued function defined on R^n, then an r-dimensional unfolding of η is a germ $f: R^n \times R^r \longrightarrow R$ of a smooth function defined near the origin, such that $f|R^n \times \{0\}$, considered as a germ on R^n, is η. An r-dimensional unfolding is in effect a germ of a smooth r-parameter family of germs on R^n; the fibration of R^{n+r} as $R^n \times R^r$ plays an important role in the theory of unfoldings; essentially it is part of the structure of an unfolding. For this reason the theory of unfoldings has important applications to Thom's catastrophe theory, since there one considers families of potential functions.

To define a notion of stability, one first needs a notion of equivalence between objects. This is usually given by defining two objects to be equivalent if one can be transformed into the other by homeomorphisms or diffeomorphisms of the underlying space. In the case of unfoldings, these diffeomorphisms will be required to respect the fibration of $R^n \times R^r$; for this reason the theory of stable unfoldings is distinct from the theory of stable germs without additional structure, and in particular is non-trivial.

We define several different notions of stability; most of these are defined geometrically, and the difference between these notions results from choosing different equivalence relations and different topologies on the space of smooth real valued functions defined on an open subset of R^{n+r}. One of the stability notions, infinitesimal stability, is defined by an algebraic condition. The main result of this paper (Theorem 4.11) is that these notions are all equivalent; i.e. there is essentially only one reasonable notion of stability for unfoldings. As a corollary we have an easily verifiable algebraic criterion for stability, since infinitesimal stability is defined by an algebraic condition.

The main application of this result is to state precisely and to prove René Thom's celebrated statement that there are exactly seven "elementary catastrophes". This is our theorem 5.6. Essentially this is a classification theorem for stable unfoldings of unfolding dimension < 4.

In the early part of the paper we investigate some aspects of the theory of germs of smooth real-valued functions and of the theory of unfoldings which are important for the investigation of stability and which also have applications in Thom's catastrophe theory. Our main reference for this part of the paper is a set of notes by John Mather [10] which exist only in manuscript form. These notes have not been published nor does it appear that they are likely to be published in the near future. Therefore we have included in this early part of the paper several of Mather's proofs, since they are not likely to be otherwise available to the reader.

The paper is organized as follows:

§ 1 contains the tools which are needed throughout the rest of the paper. Here we define our notation and recall the definitions of basic concepts. We also quote some major theorems from other sources which will be applied in our investigations, and we prove some easy corollaries of these theorems. The most important results which we cite in this chapter are: Nakayama's lemma, the Malgrange preparation theorem; the Thom transversality lemma; and a lemma of Mather's on constructing certain germs of diffeomorphisms.

§ 2 is concerned with finite determinacy of germs. We work with two equivalence relations between germs: suppose μ and η are germs at 0 of functions on $\mathbb{R}^n$; they are <u>right equivalent</u> if there is a local diffeomorphism φ of $\mathbb{R}^n$ such that $\mu = \eta\varphi$; they are <u>right-left</u> equivalent if there is a local diffeomorphism φ of $\mathbb{R}^n$ and a local diffeomorphism λ of $\mathbb{R}$ such that $\mu = \lambda\eta\varphi$. A germ is right (right-left) k-determined if it is right (right-left) equivalent to any other germ with the same k-jet. We investigate a large number of algebraic criteria for k-determinacy. Many of the results in this section were first proved by Mather [6] and [10]; however the results of [6] are weaker than those we give here and in [10] Mather considers only the "right" case. We have improved slightly upon the results of [6] and generalized the results of [10] to the "right-left" case.

§ 3 deals with unfoldings of germs. Here again there is a "right" and a "right-left" case, depending upon whether we transform unfoldings by mappings on the right only or on the right and the left. In this section we investigate the problem of determining what unfoldings a given germ can have. We give an almost complete solution to this problem by showing that every finitely determined germ has right and right-left universal unfoldings: an unfolding f of a germ η is called right (right-left) universal if it induces every other unfolding of η by composition with mappings on the right or on the right and the left. We also give algebraic characterizations of the universal unfoldings. Here too the results for the "right" case are due to Mather [10]; we have generalized these results to the "right-left" case.

§ 4 is concerned with stability of unfoldings and contains our main result. We define six geometric notions of stability, and one algebraically-defined notion, infinitesimal stability. We prove that infinitesimal stability is equivalent to right-left universality. We conclude the section with our main theorem, that all of the notions of stability we define are equivalent to each other (and hence equivalent to right-left universality).

§ 5 contains the main application of our results, to Thom's catastrophe theory. We give a precise statement of Thom's claim that there are only seven elementary catastrophes, and prove the validity of his list. The results of § 4, which establish a notion of stability for unfoldings and give an algebraic criterion for stability, are necessary for the statement of the Thom theorem and are applied in the proof. A major step in the proof is a classification theorem for germs of low codimension (Lemma 5.15) which is essentially due to Mather [10]; we give the proof of this lemma anyway, since it has not been published. We also give a new proof of the splitting lemma (our Lemma 5.12), which Mather proves in a different way, and with additional hypotheses, in [10]. Gromoll and Meyer prove a generalisation of this lemma in [2].

The paper concludes with a short appendix in which the main ideas of Thom's catastrophe theory are sketched and the relevance of the results of this paper to Thom's theory is explained.

The author would like to express his especial gratitude to Prof. Dr. Klaus Jänich for his excellent course of lectures on the theory of singularities, which inspired the production of this paper; for his invaluable support and advice; in short, for having made this work possible. Many thanks are also due to Les Lander, who acted as guinea pig for many of the formulations in the text, and to Dr. Th. Bröcker for many useful conversations and for intellectual stimulation. Finally, thanks are due to Frl. Kilger, secretary of the mathematics department at Regensburg, who performed the unthankful job of typing the manuscript.

§ 1 PRELIMINARIES

We begin by recalling some well-known definitions.

Definition 1.1. Let X and Y be topological spaces, and let $x \in X$. A map-germ $f: X \longrightarrow Y$ at x is an equivalence class of continuous mappings $g: U \longrightarrow Y$, where U is a neighbourhood of x in X, and where $g: U \longrightarrow Y$ is equivalent to $h: V \longrightarrow Y$ if and only if there is a neighbourhood $W \subset U \cap V$ of x in X such that $g|W = h|W$. If g is a member of the equivalence class f we call g a representative of f and say f is the germ at x of g.

If X and Y are smooth manifolds, then a map-germ $f: X \longrightarrow Y$ at $x \in X$ is said to be smooth or C^∞ if it has a representative which is smooth on a neighbourhood of x.

If $f: X \longrightarrow Y$ is a map-germ at $x \in X$ and $g: Y \longrightarrow Z$ is a map-germ at $f(x) \in Y$, we can define a map-germ $g \circ f: X \longrightarrow Z$ at x in the obvious way, by composing representatives of f and g and taking the germ of the result.

Similarly, by first taking representatives, we can perform arithmetic operations on germs of maps into $\mathbb{R}$.

If $Z \subseteq X$ and $x \in \bar{Z}$ we may speak of the restriction to Z of a map-germ $f: X \longrightarrow Y$ at x.

We shall say a map-germ $f: X \longrightarrow Y$ at x is constant (or 0) on X or on $Z \subseteq X$ if every representative is constant (or 0) in a neighbourhood of x, on X or on $Z \subseteq X$ respectively.

In future, we shall often perform operations such as the above on germs without comment and without specific reference to representatives. We shall mention representatives of germs explicitly only in the rare cases when it is necessary for clarity.

Definition 1.2. Let X and Y be smooth manifolds. Let $x \in X$. If k is a non-negative integer, then a k-jet $z: X \longrightarrow Y$ at x is an equivalence class of smooth mappings $g: U \longrightarrow Y$, where U is a neighbourhood of x in X, and where $g: U \longrightarrow Y$ is equivalent to $h: V \longrightarrow Y$ if and only if $g(x) = h(x)$ and all their partial derivatives of order $\leq k$ at x agree (in some, and hence in any, system of local coordinates near x in X and near $g(x) = h(x)$ in Y). If g is a member of the equivalence class z, we say z is the k-jet of g at x, and denote z by $j^k g(x)$.
Since the k-jet of g at x obviously depends only on the germ at x of g, we may also speak of the k-jet at x of a map-germ f at x. We shall use in this case the same notation $j^k f(x)$.

Remark: For simplicity, we shall often name germs and jets by giving the name of a representative. For example, 0 will denote the germ and the k-jets of the function from $X \longrightarrow R$ whose value is 0 everywhere on X.
In the remainder of this paper, we shall usually be considering germs and jets at 0 of mappings of Euclidean spaces, and we shall be using the following notations:

Definition 1.3. We denote by $\mathcal{E}(n,p)$ the set of germs at $0 \in \mathbb{R}^n$ of smooth mappings from $\mathbb{R}^n$ to $\mathbb{R}^p$. If $p = 1$, we shall write $\mathcal{E}(n)$ for $\mathcal{E}(n,1)$.

$\mathcal{E}(n,p)$ can be made into an $\mathbb{R}$-vector space in the obvious way. In fact, if $p = 1$, $\mathcal{E}(n)$ has a natural $\mathbb{R}$-algebra structure induced by the $\mathbb{R}$-algebra structure of $\mathbb{R}$. $\mathcal{E}(n)$ as a ring has a unique maximal ideal $m(n)$; $m(n)$ is the set of germs $f: \mathbb{R}^n \longrightarrow \mathbb{R}$ at 0 such that $f(0) = 0$.

If $f \in \mathcal{E}(n,p)$ and $1 \leqslant i \leqslant p$, then f_i will denote the germ $\in \mathcal{E}(n)$ of the composition $y_i \circ f$, where $y_i \in \mathcal{E}(p)$ is the germ at 0 of the i-th coordinate function of $\mathbb{R}^p$.

Let g be an element of $\mathcal{E}(n,p)$ and let $g(0) = 0$. Then for any r, the germ g induces a canonical $\mathbb{R}$-linear mapping $g^*: \mathcal{E}(p,r) \longrightarrow \mathcal{E}(n,r)$ defined by $g^*(f) = f \circ g$ if $f \in \mathcal{E}(p,r)$. If $r = 1$, then $g^*: \mathcal{E}(p) \longrightarrow \mathcal{E}(n)$ is a homomorphism of $\mathbb{R}$-algebras.

Lemma 1.4. Let $0 \leqslant r \leqslant n$ and let $i: \mathbb{R}^{n-r} \longrightarrow \mathbb{R}^n$ be the linear embedding defined by $i(t_1,\dots,t_{n-r}) = (0,\dots,0,t_1,\dots,t_{n-r})$. Let $K = i(\mathbb{R}^{n-r}) = \{(x_1,\dots,x_n) \in \mathbb{R}^n | x_1 = \dots = x_r = 0\}$. Let m denote the ideal of $\mathcal{E}(n)$ generated by $x_1,\dots,x_r$, that is, by the germs at 0 of the first r coordinate functions of $\mathbb{R}^n$.

Then the following are equivalent for $f \in \mathcal{E}(n)$:

(a) $f \in m^{k+1}$

(b) f has a representative $\underset{\sim}{f}: U \longrightarrow \mathbb{R}$ which vanishes together with all of its partial derivatives of order $\leqslant k$ everywhere on $K \cap U$.

For the proof, see [6, lemma 1.4]

In order to state the next lemma, we need some abbreviated notation.

<u>Definition 1.5.</u> Let $\alpha = (\alpha_1,\ldots,\alpha_r)$ be an r-tuple of non-negative integers. Let $x_1,\ldots,x_n$ be the coordinates of $\mathbb{R}^n$. We define

$$|\alpha| := \Sigma_{i=1}^r \alpha_i$$
$$x^\alpha := x_1^{\alpha_1} x_2^{\alpha_2} \ldots x_r^{\alpha_r}$$
$$D_\alpha = \partial^{|\alpha|}/\partial x_1^{\alpha_1} \partial x_2^{\alpha_2} \ldots \partial x_r^{\alpha_r}$$ = the "αth" partial differentiation with respect to $x_1,\ldots,x_r$.
$$\alpha! = \alpha_1!\alpha_2! \ldots \alpha_r!$$

<u>Lemma 1.6.</u> (see e.g. [1, p. 13]): Let $0 \leq r \leq n$, and let i, K and $\mathfrak{m}$ be as in lemma 1.4. Let $\pi\colon \mathbb{R}^n \longrightarrow \mathbb{R}^{n-r}$ be the canonical projection defined by $\pi(x_1,\ldots,x_n) := (x_{r+1},\ldots,x_n)$.

Now let $f \in \mathcal{E}(n)$ and let $k \geq 0$. Then there exist unique germs $g_\alpha \in \pi^*\mathcal{E}(n-r)$, where α varies over all r-tuples of non-negative integers with $|\alpha| \leq k$, and there exists a unique $h \in \mathfrak{m}^{k+1}$, such that

(a) $$f = \Sigma_{|\alpha|\leq k}\, g_\alpha \cdot x^\alpha + h\ .$$

(Note that $\pi^*\mathcal{E}(n-r)$ is just the set of all germs in $\mathcal{E}(n)$ of functions which depend only on $x_{r+1},\ldots,x_n$)

<u>Proof:</u> Define $g_\alpha = (\pi^* i^* D_\alpha f)/\alpha!$ and define $h := f - \Sigma_{|\alpha|\leq k} g_\alpha x^\alpha$. Now consider any $\beta = (\beta_1,\ldots,\beta_r)$ with

$|\beta| \leq k$. Since g_α depends only on $x_{r+1},\dots,x_n$ it follows that $\partial g_\alpha/\partial x_i = 0$ if $1 \leq i \leq r$. Hence $D_\beta g_\alpha x^\alpha = g_\alpha D_\beta x^\alpha$ for all α with $|\alpha| \leq k$. Now $D_\beta x^\alpha = \alpha!$ if $\alpha = \beta$; if $\alpha \neq \beta$ then $D_\beta x^\alpha$ is either 0 or a non-constant monomial in $x_1,\dots,x_r$. In any event, if $\alpha \neq \beta$, then $D_\beta x^\alpha$ is identically 0 on K.

So for any β with $|\beta| \leq k$ and at any point $x \in K$, we have
$D_\beta h(x) = D_\beta f(x) - \Sigma_{|\alpha|\leq k} g_\alpha(x)\cdot D_\beta x^\alpha(x) = D_\beta f(x) - \beta! g_\beta(x) = 0$
because of the way g_β was defined. Note that this implies that <u>all</u> the derivatives of h of order $\leq k$ vanish on K near 0, even if we differentiate with respect to $x_{r+1},\dots,x_n$. Hence by lemma 1.4, we have $h \in m^{k+1}$.

We have defined g_α and h so that (a) is obviously satisfied. So we have completed the existence part of the proof.

To show the uniqueness of g_α and h we apply D_β to (a) and restrict to K, for each β with $|\beta| \leq k$. On the left side of the resulting equation we get $D_\beta f|K$. For the right hand side we observe that it follows from Lemma 1.4 that $D_\beta h|K = 0$; and we can repeat our argument above to compute $D_\beta g_\alpha x^\alpha|K$. So on the right-hand side we get $\beta! g_\beta|K$. But since g_β must be in $\pi^* \mathcal{E}(n-r)$, $g_\beta|K$ determines g_β uniquely, so g_β must be as we defined it above. But then h must also be as we defined it above, and so h and the g_β are unique. qed.

We have the following easy corollaries of Lemma 1.4 and Lemma 1.6:

<u>Corollary 1.7.</u> $m(n)$ is generated by the germs $x_1, x_2, \ldots, x_n$.

<u>Corollary 1.8.</u> $m(n)^k$ consists of exactly those germs which vanish at 0 together with their derivatives of order $< k$, that is, just those germs whose $k-1$ jet at 0 vanishes.

<u>Definition 1.9.</u> We define $m(n)^\infty := \bigcap_k m(n)^k$

<u>Corollary 1.10.</u> Let M be the augmentation ideal of $\mathbb{R}[x_1, \ldots, x_n]$, generated by $x_1, \ldots, x_n$. Let m be the unique maximal ideal of $\mathbb{R}[[x_1, \ldots, x_n]]$, generated by $x_1, \ldots, x_n$.

Then for $0 \leq k < \infty$, $\mathcal{E}(n)/m(n)^k \cong \mathbb{R}[[x_1, \ldots, x_n]]/m^k$

$$\cong \mathbb{R}[x_1, \ldots, x_n]/M^k$$

and $\mathcal{E}(n)/m(n)^\infty \cong \mathbb{R}[[x_1, \ldots, x_n]]$.

The isomorphisms are given by the Taylor series expansion. Note that in the case $k = \infty$ it follows from a well-known classical result of E. Borel that every formal power series is the Taylor series at 0 of some germ, so that the isomorphism is in fact onto $\mathbb{R}[[x_1, \ldots, x_n]]$.

<u>Definition 1.11.</u> An element f of $\mathcal{E}(n)$ will be called a singularity if $f(0) = 0$ and f has a representative $\underline{f}: U \longrightarrow \mathbb{R}$ which is singular at 0 (i.e. $\partial\underline{f}/\partial x_i(0) = 0$ for all i, $1 \leq i \leq n$). This condition is equivalent to saying $j^1 f(0) = 0$, or (by 1.8) to saying $f \in m(n)^2$.

We shall introduce some algebraic notations which will be used frequently throughout this paper.

<u>Definition 1.12.</u> Let S be a commutative ring and let A be an S-module. If $a_1,\ldots,a_r$ are elements of A we shall denote the S-submodule of A generated by the a_i's by $\langle a_1,\ldots,a_r\rangle$, or sometimes, more explicitly, by $\langle a_1,\ldots,a_r\rangle_S$.

As a special case, if A = S considered as a module over itself, and if $a_1,\ldots,a_r$ are in S, we shall write $\langle a_1,\ldots,a_r\rangle$, or sometimes $\langle a_1,\ldots,a_r\rangle_S$, to denote the ideal of S generated by the a_i's.

We shall use the more explicit notation when it is necessary for clarity (for example, when several different rings act on A) and the shorter notation only when no confusion can result.

<u>Remark:</u> Frequently we shall consider a situation in which several rings act on a module, but where one of the rings plays a major role in that its action induces the action of the other rings in a natural way. To be more precise:

Let $S,T_1,\ldots,T_k$ be commutative rings, and let A be an S-module. Suppose further that for each T_j one of the following statements is true:

i) T_j is an ideal of S.

or ii) S is an R-algebra (with identity) and $T_j = R$

or iii) T_j is a subring of S.

or iv) A homomorphism $\varphi\colon T_j \longrightarrow S$ has been given, so that every S-module can be made, via φ, into a T_j-module in a natural way.

In each of these cases ((i), (ii) and (iii) are, of course, special cases of (iv)) there is a natural T_j-module structure on A induced by the S-module structure.

Then if $a_1,\ldots,a_r$ are in A we shall write $\langle a_1,\ldots,a_r\rangle$ to denote the S-submodule of A generated by the a_i's, and we shall denote the T_j-submodule generated by the a_i's by $\langle a_1,\ldots,a_r\rangle_{T_j}$.

In our applications of this convention, it will always be clear which of the rings involved plays the role of S.

Lemma 1.13. (Nakayama's Lemma). Let R be a commutative ring with identity and let I be an ideal in R such that $1+z$ is invertible for all $z \in I$. Let A and B be submodules of some R-module M and suppose A is finitely generated over R.

If (a) $B + I\cdot A \supseteq A$,

then (b) $B \supseteq A$

and if equality holds in (a), then equality holds in (b).

Proof: Suppose $B \not\supseteq A$. Now because A is finitely generated, and because $B + A \supseteq A$ but $B \not\supseteq A$, we can find elements $a_1,\ldots,a_k$ of A ($k \geq 1$) such that $B + \langle a_1,\ldots,a_k\rangle \supseteq A$ but $B + \langle a_2,\ldots,a_k\rangle \not\supseteq A$. Let $C = B + \langle a_2,\ldots,a_k\rangle$. Clearly (a) implies $C + I\cdot A \supseteq A$. Moreover $A \subseteq C + \langle a_1\rangle$, so we have $A \subseteq C + I\cdot A \subseteq C + I\cdot C + Ia_1 = C + Ia_1$. In particular, there exist $c \in C$ and $z \in I$ such that $a_1 = c + za_1$. But $1-z$ is a unit in R, so $a_1 = (1-z)^{-1}c$ and hence $a_1 \in C$. But then $C + \langle a_1\rangle = C$, and so $A \subseteq C$. This is a contradiction. So $A \subseteq B$.

If equality holds in (a), then we also have $B \subseteq A$, so $A = B$. A slightly different proof can be found in [5,p.281]

<u>Corollary 1.14.</u> (See [6, Corollary 1.6]). Let A be a finitely generated $\mathcal{E}(n)$ module and let B be a submodule of A such that for some k

$$\dim_{\mathbb{R}} A/(m(n)^{k+1}A + B) \leq k .$$

Then $m(n)^k A \subseteq B$.

<u>Proof:</u> Let $C = A/B$. Then clearly $\dim_{\mathbb{R}} C/m(n)^{k+1}C \leq k$. Now for each p, $0 \leq p \leq k$, we have $m(n)^{p+1}C \subseteq m(n)^p C$ and if the inclusion is proper we have $\dim_{\mathbb{R}} m(n)^{p+1}C/m(n)^{k+1}C < \dim_{\mathbb{R}} m(n)^p C/m(n)^{k+1}C$. So there is a p, $0 \leq p \leq k$, such that $m(n)^{p+1}C = m(n)^p C$. Now we apply Nakayama's lemma with $R = \mathcal{E}(n)$, $I = m(n)$, $A = m(n)^p C$, and $B = 0$ and conclude: $m(n)^p C = 0$. But this implies $m(n)^k A \subseteq B$.

We shall make frequent use of the Malgrange preparation theorem, which we state below in Mather's formulation ([6,p.132]), which is a slight generalisation of Malgrange's form.

<u>Theorem 1.15.</u> (Malgrange Preparation Theorem). Let $f \in \mathcal{E}(n,p)$ and let $f(0) = 0$. Let A be a finitely generated $\mathcal{E}(n)$-module and suppose $\dim_{\mathbb{R}} A/f^*(m(p))A$ is finite. Then A is finitely generated as an $\mathcal{E}(p)$ module via f^*.

For a proof, see e.g. [6,pp.131-134] (the division lemma which this proof uses is proved in [4]), or see [3,Chapter V] or see the articles of Wall, Nirenberg, Łojasiewicz, Mather and Glaeser in [18,pp.90-132].

This however is not the form in which we shall use the theorem. We shall apply the Malgrange theorem in the following formulation, which is also due to Mather ([6, lemma, p. 134]).

<u>Theorem 1.16.</u> (Malgrange preparation theorem). Let $f \in \mathcal{E}(n,p)$ and let $f(0) = 0$. Let C be a finitely generated $\mathcal{E}(n)$ module. Then C can also be considered as an $\mathcal{E}(p)$ module via f^*. Let A be a finitely generated $\mathcal{E}(p)$-submodule of C and let B be an $\mathcal{E}(n)$-submodule of C.

If (a) $A + B + f^*\mathfrak{m}(p)C = C$,

then (b) $A + B = C$.

<u>Proof:</u> Set $C' = C/B$. Let $\pi\colon C \longrightarrow C'$ be the canonical projection and set $A' = \pi(A)$. Then (a) implies

(c) $A' + f^*\mathfrak{m}(p)C' = C'$.

Now A' is an $\mathcal{E}(p)$ module, and if we consider C' as an $\mathcal{E}(p)$ module via f^*, we may rewrite (c) as

(d) $A' + \mathfrak{m}(p)C' = C'$.

Since A' is finitely generated over $\mathcal{E}(p)$, it follows that $C'/\mathfrak{m}(p)C'$ is finitely generated over $\mathcal{E}(p)$ and hence is also finite-dimensional as a vector space over $\mathbb{R}$. But $C'/\mathfrak{m}(p)C'$ is just $C'/f^*\mathfrak{m}(p)C'$, and since C' is finitely generated over $\mathcal{E}(n)$, Theorem 1.15 tells us C' is finitely generated over $\mathcal{E}(p)$. But then from (d) it follows by Nakayama's lemma that $A' = C'$. But this clearly implies that $C = A + B$.

The next theorem is a slight generalization of this result.

Theorem 1.17. (see [6, Th. 1.13]). Let $f \in \mathcal{E}(n,p)$ and let $f(0) = 0$. Let C be a finitely-generated $\mathcal{E}(n)$-module. C is also a module over $\mathcal{E}(p)$ via f^*. Let B be an $\mathcal{E}(n)$-submodule of C and let A be a finitely generated $\mathcal{E}(p)$-submodule of C. Let $k = \dim_R A/m(p)A$.

Then

(a) $$A + B + (f^*m(p) + m(n)^{k+1})C = C$$

implies

(b) $$A + B = C .$$

Proof: Set $B' = B + f^*m(p)C$. Since A is an $\mathcal{E}(p)$-submodule of C, we have $m(p)A \subseteq f^*m(p)C \subseteq B'$. Hence from (a) it follows that $\dim_R C/(B' + m(n)^{k+1}C) \leq \dim_R A/m(p)A = k$. So by Corollary 1.14, $m(n)^k C \subseteq B'$. So from (a) it follows that $A + B + f^*m(p)C = C$. But then Theorem 1.16 implies $A + B = C$.

Lemma 1.18. Let A be a cyclic $\mathcal{E}(1)$ module. Then for any non-negative integer k, $m(1)^k A$ is a cyclic submodule of A and $\dim_R A/m(1)^k A \leq k$.
Moreover if B is any $\mathcal{E}(1)$ submodule of A and if $\dim_R A/B = r < \infty$, then $B = m(1)^r A$ (and hence is cyclic). If $\dim_R A/B = \infty$, then $B \subseteq m(1)^\infty A$.
(Note: A cyclic module is one which is generated by a single element).

Proof: Let a be a generator of A and let x be the canonical generator of $\mathfrak{m}(1)$. Then $\mathfrak{m}(1)^k$ is generated over $\mathcal{E}(1)$ by x^k, so $\mathfrak{m}(1)^k A$ is generated over $\mathcal{E}(1)$ by $x^k a$ and hence is cyclic. Moreover $a, xa, \ldots, x^{k-1}a$ generate $A/\mathfrak{m}(1)^k A$ over $\mathbb{R}$, so $\dim_{\mathbb{R}} A/\mathfrak{m}(1)^k A \leq k$.

Now let B be a submodule of A. Suppose $\dim_{\mathbb{R}} A/B = r < \infty$. Then by corollary 1.14 $\mathfrak{m}(1)^r A \subseteq B$. But $\dim_{\mathbb{R}} A/\mathfrak{m}(1)^r A \leq r = \dim_{\mathbb{R}} A/B$, so the inclusion cannot be proper. Hence $\mathfrak{m}(1)^r A = B$.

Now suppose $\dim_{\mathbb{R}} A/B = \infty$. Suppose $B \not\subseteq \mathfrak{m}(1)^\infty A$. Then for some element $b \in B$ there is a largest integer s such that $b \in \mathfrak{m}(1)^s A$. But then there is a real number $c \neq 0$ and an element $\omega \in \mathfrak{m}(1)$ such that $b = (cx^s + x^s \omega(x))a$. Since $c + \omega(x)$ is a unit in $\mathcal{E}(1)$ it follows that $x^s a \in B$ and hence $\mathfrak{m}(1)^s A \subseteq B$. But this implies $\dim_{\mathbb{R}} A/B$ is finite, which is a contradiction. So $B \subseteq \mathfrak{m}(1)^\infty A$.

Theorem 1.19. Let $f \in \mathfrak{m}(n)$. Let C be a finitely-generated $\mathcal{E}(n)$ module. Then C is also an $\mathcal{E}(1)$-module via f^*. Let A be a cyclic $\mathcal{E}(1)$-submodule of C and let B be an $\mathcal{E}(n)$-submodule of C. Let D be an $\mathcal{E}(n)$-submodule of C such that $\dim_{\mathbb{R}} C/D$ is finite.

If (a) $A + B + (f^*\mathfrak{m}(1) + \mathfrak{m}(n)^2)D \supseteq D$

then (b) $A + B \supseteq D$.

Moreover, if in addition $f \in \mathfrak{m}(n)^2$, then (c) $\mathfrak{m}(n)D \subseteq B$.

Proof: Let $D' = B + D$ and let $A' = A \cap D'$. Then (a) implies

(d) $$A' + B + (f^*m(1) + m(n)^2)D' = D' .$$

Now let $k = \dim_R C/D$. Then $\dim_R C/D' \leq k$ so by corollary 1.14, $m(n)^k C \subseteq D'$. Now since C is finitely generated over $\mathcal{E}(n)$, $\dim_R C/m(n)^k C$ is finite, so $\dim_R D'/m(n)^k C$ is also finite. Moreover $m(n)^k C$ is finitely generated over $\mathcal{E}(n)$, so it follows that D' is finitely generated over $\mathcal{E}(n)$. Now A' is an $\mathcal{E}(1)$-submodule of A and $\dim_R A/A' \leq \dim_R C/D' \leq k < \infty$. So by Lemma 1.18, A' is a cyclic $\mathcal{E}(1)$-module and hence $\dim_R A'/m(1)A' \leq 1$. So we can apply Theorem 1.17 to (d) to conclude that

(e) $A' + B = D'$. But this implies $A + B \supseteq D$.

Now suppose further that $f \in m(n)^2$. Then $f^*m(1) \subseteq m(n)^2$. Moreover (e) implies $A' + B + f^*m(1)D' = D'$, and since A' is an $\mathcal{E}(1)$ submodule of D' we have $m(1)A' \subseteq f^*m(1)D'$, so $\dim_R D'/(B + f^*m(1)D') \leq \dim_R A'/m(1)A' \leq 1$. So by corollary 1.14, $m(n)D' \subseteq B + f^*m(1)D' \subseteq B + m(n)^2 D'$. But then by Nakayama's Lemma (Lemma 1.13), we have $m(n)D' \subseteq B$, and hence (c) holds.

Mather proves a similar result, with more general hypotheses but with a weaker conclusion, in [6, Corollary 1.14].

We shall need some more notations for dealing with jets.

Definition 1.20. Let k be a non-negative integer. We denote by $J^k(n,p)$ the set of all k-jets at 0 of germs in $\mathcal{E}(n,p)$.

If f and g are in $\mathcal{E}(n,p)$, then f and g have the same k-jet at 0 if and only if $f_i - g_i \in m(n)^{k+1}$ for all i, $1 \leq i \leq p$. So there is a natural bijection between $J^k(n,p)$ and $(\mathcal{E}(n)/m(n)^{k+1}) \times \ldots \times (\mathcal{E}(n)/m(n)^{k+1})$ (p factors), which with the usual operations (defined coordinatewise) is an $\mathbb{R}$-algebra, and is isomorphic to $(\mathbb{R}[x_1,\ldots,x_n]/M^{k+1}) \oplus \ldots \oplus (\mathbb{R}[x_1,\ldots,x_n]/M^{k+1})$ (p summands) if k is finite. (Here M is the augmentation ideal of $\mathbb{R}[x_1,\ldots,x_n]$, generated by the x_i). This bijection defines a canonical $\mathbb{R}$-algebra structure on $J^k(n,p)$, and in fact the algebraic operations on $J^k(n,p)$ are induced by the corresponding operations on representatives of the jets. Moreover the preceding discussion shows that as an $\mathbb{R}$-vector space $J^k(n,p)$ is isomorphic to $\mathbb{R}^s$, where $s = p\binom{n+k}{k}$. These isomorphisms induce a canonical smooth differentiable structure on $J^k(n,p)$. As a standard basis of the real vector space $J^k(n,p)$ we may take the set of k-jets at 0 of all germs $f \in \mathcal{E}(n,p)$ such that for some i $(1 \leq i \leq p)$, f_i is a monomial of degree $\leq k$ in the coordinates of $\mathbb{R}^n$, and $f_j = 0$ if $j \neq i$. (So if $p = 1$ the standard basis consists of the k-jets at 0 of all monomials of degree $\leq k$ in $x_1,\ldots,x_n$). Using this standard basis, we may identify any element of $J^k(n,p)$ in a standard way with a unique p-tuple of polynomials of degree $\leq k$ in $x_1,\ldots,x_n$.

Let $k \leq r$. Then clearly two germs which have the same r-jet at 0 will also have the same k-jet at 0, so there is a natural projection $\pi_{r,k}\colon J^r(n,p) \longrightarrow J^k(n,p)$, defined by forgetting the higher-order terms. Under the isomorphisms defined in the preceding paragraph the $\pi_{r,k}$ correspond to linear projections of Euclidean spaces. We also have for each k a canonical projection π_k: $\mathcal{E}(n,p) \longrightarrow J^k(n,p)$ which assigns to each germ its k-jet at 0. Obviously, if $k \leq r$ we have $\pi_k = \pi_{r,k}\pi_r$.

We make the following observation: If $f \in \mathcal{E}(n,p)$ and $g \in \mathcal{E}(p,r)$ and if $f(0) = 0$, then the k-jet at 0 of the composition $g \circ f \in \mathcal{E}(n,r)$ depends only upon the k-jet at 0 of f and of g. In particular this means that if $f \in \mathcal{E}(n,p)$ and if $f(0) = 0$, then for each k the germ f induces a well-defined $\mathbb{R}$-linear map ${}_kf^*$: $J^k(p,r) \longrightarrow J^k(n,r)$ such that if $g \in \mathcal{E}(p,r)$, then ${}_kf^*\pi_k(g) = \pi_k(g \circ f)$. Similarly if $z \in J^k(n,p)$ and if $\pi_{k,0}(z) = 0$, then z induces a well-defined $\mathbb{R}$-linear map z^*: $J^k(p,r) \longrightarrow J^k(n,r)$, defined by setting $z^* := {}_kf^*$, where $f \in \mathcal{E}(n,p)$ is any representative of z.

Let U be an open subset of $\mathbb{R}^n$, and let f: $U \longrightarrow \mathbb{R}^p$ be a smooth mapping. We define for each k a smooth mapping J^kf: $U \longrightarrow J^k(n,p)$, called the k-jet section of f, as follows: For each $x \in U$ define a germ $f_x \in \mathcal{E}(n,p)$ by setting $f_x(z) := f(x+z)$ for all z near $0 \in \mathbb{R}^n$. Now define J^kf by letting $J^kf(x)$ be the k-jet at 0 of f_x .

Remark: At the beginning of this chapter we defined $j^kf(x)$ to be the k-jet of f at x. Note that unless $x = 0$ this is not the same thing as $J^kf(x)$, since $J^kf(x)$ is always a jet at 0.

From the definition of $J^k f$ it is clear that the germ at $x \in U$ of $J^k f$ depends only on the germ at x of f. Hence for each k we can in the obvious way associate to every germ $g \in \mathcal{E}(n,p)$ a germ $J^k g$ which is the germ at 0 of a smooth mapping from $\mathbb{R}^n$ into $J^k(n,p)$.

It will often be convenient to ignore the zero-order part of a jet, that is, the actual value of the representative function at the point where the jet is taken. For this reason we introduce the following notations:

Definition 1.21. For each non-negative integer k, define

$$J_0^k(n,p) := \{z \in J^k(n,p) \mid \pi_{k,0}(z) = 0\} .$$

$J_0^k(n,p)$ is just the set of k-jets at 0 of germs in $\mathcal{E}(n,p)$ whose zero-jet is 0, that is, whose value at $0 \in \mathbb{R}^n$ is 0. $J_0^k(n,p)$ is a linear subspace of $J^k(n,p)$ of codimension p, and so also has the structure of a finite-dimensional smooth manifold. Note that we may choose as a standard basis for $J_0^k(n,p)$ the obvious subset of our standard basis for $J^k(n,p)$.

For each k there is a canonical projection $\rho_k: J^k(n,p) \longrightarrow J_0^k(n,p)$ defined by $\rho_k j^k f(0) = j^k(f-f(0))(0)$. ρ_k just forgets the zero-order part of the jet.

Recall that if $f \in \mathcal{E}(n,p)$ and if $f(0) = 0$, then f induces $\mathbb{R}$-linear maps ${}_k f^*: J^k(p,r) \longrightarrow J^k(n,r)$. Observe that ${}_k f^*(J_0^k(p,r)) \subseteq J_0^k(n,r)$. Similarly, we remarked that if $z \in J_0^k(n,p)$, then z induces an $\mathbb{R}$-linear map $z^*: J^k(p,r) \longrightarrow J^k(n,r)$. Observe also that $z^*(J_0^k(p,r)) \subseteq J_0^k(n,r)$.

Finally, if U is an open neighbourhood of $0 \in R^n$ and if $f: U \to R^p$ is a smooth mapping, we can define for each k a smooth mapping $J_o^k f: U \to J_o^k(n,p)$ by setting $J_o^k f = \rho_k J^k f$. Similarly, to every germ $g \in \mathcal{E}(n,p)$ we can associate a germ $J_o^k g$, the germ at 0 of a smooth mapping from R^n to $J_o^k(n,p)$, by setting $J_o^k g = \rho_k J^k g$.

An important result about jet sections which we shall need is Thom's celebrated transversality lemma, which we state in a rather simple form which will be sufficient for our purposes.

Theorem 1.22 (Thom's transversality lemma). Let U be an open set of R^n. Let k be a non-negative integer and let N be a smooth submanifold of $J^k(n,p)$. Let A be the set of all smooth mappings $f: U \to R^p$ such that $J^k f$ is transversal to N everywhere in U. Then A is dense in $C^\infty(U,R^p)$, the space of all smooth mappings from U into R^p.

(Remark: In this book we shall be using the weak C^∞-topology on $C^\infty(U,R^p)$. A basis for this topology is given by taking all sets $V(L,r,g,W) := \{h \in C^\infty(U,R^p) \mid J^r(g-h)(L) \subseteq W\}$, where L is any compact subset of U, r is any non-negative integer, W is any open neighbourhood of 0 in $J^r(n,p)$, and g is any element of $C^\infty(U,R^p)$. With this topology $C^\infty(U,R^p)$ is a Baire space.

Note: The weak C^∞-topology on $C^\infty(U,R^p)$ is different from the Whitney topology W_∞, which is often found in the literature (for example in Mather [5] and [8]); the Whitney topology is finer. Actually, Thom's transversality lemma (and corollary 1.23 below) hold even in the Whitney topology, although we shall not prove this here.)

The proof of Theorem 1.22 that we give is due to Boardman [1] (he actually proves it for the Whitney topology); it was unfortunately omitted from the published version of this paper. Thom's original proof ([13]; see also the paper by Levine in [18]) is much longer and more complicated. Mather proves a version of the transversality lemma (for the Whitney topology) in [8]; the idea is essentially the same as that of Boardman's proof.

Proof of Theorem 1.22: We denote by $x_1,\dots,x_n$ the coordinates of R^n. Let P be the set of all p-tuples of polynomials of degree $\leq k$ in $x_1,\dots,x_n$; we shall consider the elements of P as functions mapping R^n to R^p. Since P is an R-vector space of dimension $s = p\binom{n+k}{n}$, we may identify P with R^s and hence consider P to be an s-dimensional manifold.

If $t = (t_1,\dots,t_n)$ is any point of R^n, then the map $\alpha_t\colon P \to J^k(n,p)$ defined by $\alpha_t(f) := J^k f(t)$ is a linear isomorphism, as one easily sees.

Now let $h \in C^\infty(U,R^p)$. Consider the smooth map $\gamma\colon U \times P \to J^k(n,p)$ given by $\gamma(x,f) = J^k(h+f|U)(x)$. For each $x \in U$ clearly $\gamma|x \times P$ is a diffeomorphism onto $J^k(n,p)$, since for $f \in P$ we have $\gamma(x,f) = J^k h(x) + \alpha_x(f)$. Hence γ is a submersion. Let $M = \gamma^{-1}(N)$; then M is a submanifold of $U \times P$.

Let $\pi\colon U \times P \to P$ be the projection onto the second factor. If $f \in P$, let $i_f\colon U \times f \to U \times P$ be the inclusion. Now for $(x,f) \in U \times P$, clearly $J^k(h+f|U)$ is transversal to N at x if and only if γi_f is transversal to N at (x,f), and for this to be so it suffices that i_f be transversal to M at (x,f), since γ is a submersion and $M = \gamma^{-1}(N)$.

But clearly i_f is transversal to M at (x,f) if and only if (x,f) $\notin$ M or $\pi|M$ is regular at (x,f). Hence if $J^k(h+f|U)$ fails to be transversal to N at some x $\in$ U, then f must be a critical value of $\pi|M$. By the Sard-Brown theorem, the set of critical values of $\pi|M$ is a set of measure 0 in P. Therefore there are functions f in P, arbitrarily close to 0, such that $J^k(h+f|U)$ is transversal to N on U. But this clearly implies there are functions g arbitrarily close to h in $C^\infty(U,R^p)$ such that $J^k g$ is transversal to N on U (for if K $\subseteq$ U is compact and if $f = (f_1,\dots,f_p) \in P$, then all partial derivatives of the f_i of order $\leq$ k will be small on K if f is close to 0 in P, and the partial derivatives of the f_i of order > k vanish everywhere in R^n. Hence if f is near 0 in P, then f will also be close to 0 in $C^\infty(U,R^p)$.) qed.

Corollary 1.23. Let U be an open subset of R^n. Let k be a non-negative integer and let N be a smooth submanifold of $J^k(n,p)$. Let A be the set of all smooth mappings f: U $\to R^p$ such that $J^k f$ is transversal to N on U. Then A is a countable intersection of open dense subsets of $C^\infty(U,R^p)$. In particular, since $C^\infty(U,R^p)$ is a Baire space, A is dense.

Remark: Since an immersed submanifold of $J^k(n,p)$ can be written as a countable union of imbedded submanifolds, it is clear that the corollary also holds in the case when N is merely immersed in $J^k(n,p)$ (for example, if N is an orbit under the action of a (non-compact) Lie group). In fact it is this case for which we shall wish to apply this result.

Proof of Corollary 1.23 If C is a subset of U and D is a subset of N, we shall let A(C,D) denote the set of all smooth mappings $f: U \to R^p$ such that $J^k f$ is transversal to N on $C \cap (J^k f)^{-1}(D)$; and we shall write A(C) for A(C,N), so that A(C) is the set of all smooth mappings $f: U \to R^p$ such that $J^k f$ is transversal to N on C. We wish to show that A(U) is a countable intersection of open dense subsets of $C^\infty(U,R^p)$. By theorem 1.22 we already know A(U) is dense, so it is enough to show A(U) is a countable intersection of open sets.

Since U can be written as a countable union of compact subsets, A(U) = A is a countable intersection of sets A(K) for K compact, so it is enough to show that for any compact subset K of U, A(K) is a countable intersection of open subsets of $C^\infty(U,R^p)$.

So let $K \subseteq U$ be compact. A(K) = A(K,N) and N can be written as a countable union of compact sets, so A(K) can be written as a countable intersection of sets A(K,L) for $L \subseteq N$ compact. Hence to complete the proof it is enough to show that A(K,L) is open in $C^\infty(U,R^p)$ for any compact subset L of N. We shall prove this below.

Notation: In general, if M is any C^∞-manifold and if $x \in M$ we shall denote by $T_x M$ the tangent space to M at x. And if $g: M \to M'$ is a smooth mapping between manifolds, and $x \in M$, we denote by $T_x g : T_x M \to T_{g(x)} M'$ the differential of g at x. It is a linear mapping of the tangent spaces.

Now recall that we may identify $J^k(n,p)$ with the Euclidean space R^s, where $s = p\binom{n+k}{k}$. Hence for any $z \in J^k(n,p)$ we may also identify $T_z J^k(n,p)$ with R^s in a canonical way (depending only on the isomorphism we have chosen between $J^k(n,p)$ and R^s). Let r be the dimension of N. Then for any $z \in N$, the tangent space $T_z N$ is an r-dimensional subspace of R^s and we have a continuous map τ from N to the Grassman manifold $G_{r,s}$ of r-dimensional linear subspaces of R^s; the map τ is given by $\tau(z) = T_z N$.

Let $x_1,\dots,x_n$ be coordinates on U. If $h: U \to R^p$ is a smooth mapping and if $x \in U$, then for $1 \le i \le n$ we denote by $v_i(h,x)$ the vector $\frac{\partial J^k h}{\partial x_i}(x) = T_x J^k h\left(\frac{\partial}{\partial x_i}\right) \in T_{J^k h(x)} J^k(n,p)$; we consider $v_i(h,x)$ as a vector in R^s in accordance with the identification made above. Now $T_x J^k h(T_x U)$ is a linear subspace of R^s generated by $v_1(h,x),\dots,v_n(h,x)$. From the definition it is clear that $v_i(h,x)$ depends only on $J^{k+1}h(x)$, and moreover depends linearly on it. Obviously, therefore, $v_i(h,x)$ depends continuously on $h \in C^\infty(U,R^p)$ and $x \in U$.

Let $P = \{(F,v_1,\dots,v_n) \mid F \in G_{r,s};\ v_i \in R^s\ (i=1,\dots,n)$; and the space generated by $v_1,\dots,v_n$ is transversal to $F\}$.

Clearly P is an open subset of $G_{r,s} \times \underbrace{R^s \times \dots \times R^s}_{n}$, and $J^k h$ is transversal to N at x if and only if $J^k h(x) \notin N$ or $(\tau(J^k h(x)), v_1(h,x),\dots,v_n(h,x)) \in P$.

Now let L be a compact subset of N and let $f \in A(K,L)$. For every $x \in K$, we shall define an open neighbourhood V_x of x in U and a neighbourhood W_x of f in $C^\infty(U,R^p)$ such that $W_x \subseteq A(V_x,L)$. This will be enough to complete the proof, for K is covered by finitely many of the V_x's, say $V_{x_1},\dots,V_{x_q}$, and if we let $W = W_{x_1} \cap \dots \cap W_{x_q}$ then W is a neighbourhood of f in $C^\infty(U,R^p)$, and $W \subseteq A(K,L)$, for if $g \in W$ then $g \in A(V_{x_i},L)$ for each i, $1 \le i \le q$, and since the V_{x_i} cover K it follows that $g \in A(K,L)$.

Let $x \in K$, and let $y = J^k f(x)$.

If $y \in L$ then since $f \in A(K,L)$ we have $J^k f$ is transversal to N at x. Hence $(\tau(y), v_1(f,x),\ldots,v_n(f,x)) \in P$. Since P is open, and since τ and the v_i are continuous, there is a neighbourhood Y of y in $J^k(n,p)$, a neighbourhood X of x in U and a neighbourhood Z of f in $C^\infty(U,R^p)$ such that for $y' \in N \cap Y$, $x' \in X$ and $f' \in Z$ we have $(\tau(y'), v_1(f',x'),\ldots,v_n(f',x')) \in P$. Clearly for $h \in C^\infty(U,R^p)$ and $z \in U$ the jet $J^k h(z)$ depends continuously on h and z, so there is a neighbourhood S of $x \in U$ and a neighbourhood T of $f \in C^\infty(U,R^p)$ such that for $x' \in S$ and $f' \in T$ we have $J^k f'(x') \in Y$. Hence for $x' \in S \cap X$ and $f' \in T \cap Z$ it is clear that $J^k f'$ is transversal to N at x'. Let $V_x = S \cap X$ and let $W_x = T \cap Z$. Then clearly V_x and W_x have the desired properties; in fact $W_x \subseteq A(V_x) \subseteq A(V_x,L)$.

On the other hand, suppose $y = J^k f(x) \notin L$. Since L is compact, it is closed in $J^k(n,p)$, and hence there is a neighbourhood V_x of $x \in U$ and a neighbourhood W_x of $f \in C^\infty(U,R^p)$ such that for $x' \in V_x$ and $f' \in W_x$ we have $J^k f'(x') \notin L$. Then for $f' \in W_x$ we have $V_x \cap (J^k f')^{-1}(L) = \emptyset$, so clearly $W_x \subseteq A(V_x,L)$. This completes the proof.

In the following chapters we shall be investigating equivalence classes of germs and jets defined by the action of groups of local diffeomorphisms, so we need some notations for these groups.

<u>Definition 1.24.</u> We define $L(n) := \{\varphi \in \mathcal{E}(n,n) \mid \varphi(0) = 0$ and φ is non-singular at $0\}$. Note that whether or not a germ $\varphi \in \mathcal{E}(n,n)$ belongs to $L(n)$ depends only on the 1-jet of φ at 0.

We can make $L(n)$ into a group by taking as the group operation the composition of germs in $\mathcal{E}(n,n)$. The group $L(n)$ is the group of germs of local diffeomorphisms of $\mathbb{R}^n$ at 0.

$L(n)$ acts on $\mathcal{E}(n)$ on the right if we define, for $f \in \mathcal{E}(n)$ and $\varphi \in L(n)$, the product $f \cdot \varphi \in \mathcal{E}(n)$ to be the germ at 0 of the composition $f \circ \varphi \colon \mathbb{R}^n \longrightarrow \mathbb{R}$. Similarly $L(1)$ acts on $m(n)$ on the left if we define, for $\psi \in L(1)$ and $f \in m(n)$, the product $\psi \cdot f \in m(n)$ to be germ at 0 of the composition $\psi \circ f \colon \mathbb{R}^n \longrightarrow \mathbb{R}$. We can combine these two actions to define an action of $L(1) \times L(n)$ on $m(n)$ "on both sides"; formally we can write this action as an action from the right if we define $f \circ (\psi, \varphi) := \psi^{-1} \circ f \circ \varphi$ for $f \in m(n)$, $\varphi \in L(n)$ and $\psi \in L(1)$. Note that $L(n)$ acts on $\mathcal{E}(n)$ via $\mathbb{R}$-algebra automorphisms of $\mathcal{E}(n)$; moreover the ideals $m(n)^k$ $(1 \leq k \leq \infty)$ are invariant under the actions of <u>all three</u> groups.
$L(n)$ and $L(1)$ may be considered as subgroups of $L(1) \times L(n)$ in the obvious way.

We shall in future be interested primarily in the groups $L(n)$ and $L(1) \times L(n)$. In fact, most of the definitions in the coming chapters will refer to the actions of these two groups, so that each concept defined will have two interpretations, depending on which group action is involved.

To distinguish these two interpretations, a concept defined by reference to the group L(n), acting on $\mathcal{E}(n)$ from the right, will be called a "right-concept" (abbreviated "r-concept"); the corresponding concept defined by reference to the action of $L(1) \times L(n)$ will be called the "two-sided" or "right-left" version of the concept and will be abbreviated "rl-concept". If we omit the modifier "right" or "right-left", then either interpretation is understood.

We recall that if $f \in \mathcal{E}(n,p)$ and $g \in \mathcal{E}(p,r)$ and if $f(0) = 0$, then the k-jet at 0 of the composition $g \circ f \in \mathcal{E}(n,r)$ depends only on the k-jet at 0 of f and of g. Hence it makes sense to apply the preceding definition (1.24) also to jets:

Definition 1.25 If k is a non-negative integer, we define

$$L^k(n) := \{z \in J^k(n,n) \mid z \text{ is the k-jet at 0 of an element of } L(n)\}.$$

We remark that because membership in L(n) depends only on the 1-jet of a germ, then if $k \geq 1$ any representative in $\mathcal{E}(n,n)$ of an element of $L^k(n)$ belongs to L(n).

Now by our observation above, the group operation of L(n) induces for each k a well-defined group-operation on $L^k(n)$, so $L^k(n)$ is a group. And again, by the same observation, we have well-defined group actions of $L^k(n)$, acting on $J^k(n,1)$ on the right, and of $L^k(1) \times L^k(n)$ acting on "both sides" on $J_0^k(n,1)$, induced by the actions of L(n) on $\mathcal{E}(n)$ and of $L(1) \times L(n)$ on $\mathfrak{m}(n)$. Moreover $J_0^k(n,1)$ is invariant under the action of $L^k(n)$ on $J^k(n,1)$, so $L^k(n)$ acts also on $J_0^k(n,1)$. As for germs, so for jets also we shall speak of "right" and "right-left" concepts in coming chapters.

The groups $L^k(n)$ are open subsets of $J_0^k(n,n)$ and hence have a natural C^∞ differentiable structure. Because of the way the group operations and the group actions were defined, one easily sees that with this differentiable structure the groups $L^k(n)$ and $L^k(1) \times L^k(n)$ are Lie groups and the group actions defined above are Lie-group actions, i.e. smooth.

We may consider $L^k(n)$ as a Lie subgroup of $L^k(1) \times L^k(n)$ by identifying $L^k(n)$ with the subgroup $Id_{\mathbb{R}} \times L^k(n)$ of $L^k(1) \times L^k(n)$.

Note that the group $L^1(n)$ is isomorphic to $GL(n;\mathbb{R})$ and its action on $J_0^1(n,1) \cong \mathbb{R}^n$ corresponds under these isomorphisms to the standard right action of $GL(n,\mathbb{R})$ on $\mathbb{R}^n$.

<u>Definition 1.26.</u> Let $z \in J_0^k(n,1)$. We denote by $zL^k(n)$ the $L^k(n)$-orbit of z and by $L^k(1)zL^k(n)$ the $L^k(1) \times L^k(n)$-orbit of z.

Since $J_0^k(n,1)$ is canonically isomorphic to a Euclidean space we may identify the tangent space of $J_0^k(n,1)$ at any of its points with $J_0^k(n,1)$ itself. We shall always make this identification. Now if $z \in J_0^k(n,1)$, and if A is any 1-1 immersed submanifold of $J_0^k(n,1)$ containing z, we shall denote by

T_zA the tangent space to A at z, considered via the above identification as a linear subspace of $J_0^k(n,1)$. We shall apply this notation in particular in the cases $A = zL^k(n)$ and $A = L^k(1)zL^k(n)$.

We conclude this chapter with a lemma which we shall find useful in proving the existence of certain local diffeomorphisms.

<u>Lemma 1.27.</u> (see [6, p.144]). Let $F \in \mathcal{E}(n+1,p)$ and let $F(0) = 0$. Suppose there exist germs $\xi \in \mathcal{E}(n+1,n)$ and $\eta \in \mathcal{E}(p+1,p)$ such that for each i, $1 \leq i \leq p$, and for any x near 0 in $\mathbb{R}^n$ and any t near 0 in $\mathbb{R}$, the following equation holds:

$$\text{(a):}\quad \frac{\partial F_i(x,t)}{\partial t} = \sum_{j=1}^{n} \frac{\partial F_i(x,t)}{\partial x_j}\, \xi_j(x,t) + \eta_i(F(x,t),t).$$

Then there exist germs $\varphi \in \mathcal{E}(n+1,n)$ and $\lambda \in \mathcal{E}(p+1,p)$ such that for x near 0 in $\mathbb{R}^n$ and y near 0 in $\mathbb{R}^p$

$$\text{(b):}\quad \varphi(x,0) = x \quad \text{and} \quad \lambda(y,0) = y$$

and such that for t near 0 in $\mathbb{R}$ and x near 0 in $\mathbb{R}^n$, we have:

$$\text{(c):}\quad F(\varphi(x,t),t) = \lambda(F(x,0),t)\ .$$

Moreover, we may choose φ and λ (in fact, uniquely) such that

$$\text{(d):}\quad \frac{\partial \varphi_j(x,t)}{\partial t} = -\xi_j(\varphi(x,t),t) \quad (j = 1,\dots,n) \quad \text{and}$$

$$\frac{\partial \lambda_i(y,t)}{\partial t} = \eta_i(\lambda(y,t),t) \quad (i = 1,\dots,p)$$

for t in $\mathbb{R}$, x in $\mathbb{R}^n$, and y in $\mathbb{R}^p$.

Remark: By first choosing representatives for φ and λ we can for t near 0 in $\mathbb{R}$ define germs $\varphi_t \in \mathcal{E}(n,n)$ and $\lambda_t \in \mathcal{E}(p,p)$ by setting

$$\varphi_t(x) := \varphi(x,t) \qquad \text{for } x \in \mathbb{R}^n$$
$$\lambda_t(y) := \lambda(y,t) \qquad \text{for } y \in \mathbb{R}^p;$$

of course for $t \neq 0$ these depend on the choice of representatives but the germs at $t = 0$ of the maps $t \longrightarrow \varphi_t$ and $t \longrightarrow \lambda_t$ are well-defined.

Then observe that equations (b) imply that if t is close enough to 0, then φ_t and λ_t are germs of local diffeomorphisms of $\mathbb{R}^n$ and $\mathbb{R}^p$ respectively.

Moreover, if we define $F_t \in \mathcal{E}(n,p)$ in a similar way, that is, if, for t near 0 in $\mathbb{R}$, we set $F_t(x) := F(x,t)$ for all x near 0 in $\mathbb{R}^n$, then we can rewrite equation (c) to read:

(c') $\lambda_t^{-1} \circ F_t \circ \varphi_t = F_o$ for all t sufficiently near 0 in $\mathbb{R}$.

It is easy to see that $\lambda_t^{-1}(y)$ depends smoothly on t and y near 0. It will often be more convenient to use this notation rather than the notation used in the statement of the lemma.

Proof of lemma 1.27. Suppose that germs ξ and η have been given so that equation (a) holds. Now it follows from the fundamental existence theorem for solutions of ordinary differential equations that there are unique smooth germs $\varphi \in \mathcal{E}(n+1,n)$ and $\lambda \in \mathcal{E}(p+1,p)$ which solve the differential equations (d) and which satisfy the initial conditions (b). So we need only show that equation (c) is satisfied for these germs φ and λ.

Now we introduce the notation defined in the remark above. By this remark, since equation (b) holds, λ_t is invertible for t close enough to 0 and it suffices to prove (c'). Now if $t = 0$, equations(c) and(c') hold trivially, since φ_o and λ_o are the germs of the identity mappings of $\mathbb{R}^n$ and $\mathbb{R}^p$ respectively. So we differentiate equation(c') with respect to t; if we can show that the "derivative" of this equation holds for all t near enough to 0 we are obviously done.

So we must prove that for any t near enough to 0 in $\mathbb{R}$ and for x near 0 in $\mathbb{R}^n$,

$$(e): \qquad \frac{\partial}{\partial t}(\lambda_t^{-1} \circ F_t \circ \varphi_t)(x) = \frac{\partial}{\partial t} F_o(x) = 0 .$$

We evaluate the left-hand side:

$$(f): \qquad \frac{\partial}{\partial t}(\lambda_t^{-1} \circ F_t \circ \varphi_t)(x) = \frac{\partial\lambda_t^{-1}}{\partial t}(F(\varphi(x,t),t))$$
$$+ \sum_{i=1}^{p}\Big(\frac{\partial\lambda_t^{-1}}{\partial y_i}(F(\varphi(x,t),t))\Big[\frac{\partial F_i}{\partial t}(\varphi(x,t),t)$$
$$+ \sum_{j=1}^{n}\frac{\partial F_i}{\partial x_j}(\varphi(x,t),t)\frac{\partial\varphi_j}{\partial t}(x,t)\Big]\Big) .$$

Now we wish to evaluate $\frac{\partial\lambda_t^{-1}}{\partial t}(F(\varphi(x,t),t))$. Clearly for all y near 0 in $\mathbb{R}^p$ and t near 0 in $\mathbb{R}$ we have

$$\lambda_t^{-1}(\lambda_t(y)) = y .$$

Differentiating this equation with respect to t, we get:

$$\frac{\partial\lambda_t^{-1}}{\partial t}(\lambda_t(y)) + \sum_{i=1}^{p} \frac{\partial\lambda_t^{-1}}{\partial y_i}(\lambda_t(y))\,\frac{\partial\lambda_i}{\partial t}(y,t) = 0$$

for y and t near 0. Substituting $\lambda_t^{-1}(F(\varphi(x,t),t))$ for y in this equation, we find:

$$\frac{\partial\lambda_t^{-1}}{\partial t}(F(\varphi(x,t),t) = -\sum_{i=1}^{p} \frac{\partial\lambda_t^{-1}}{\partial y_i}(F(\varphi(x,t),t)\cdot\frac{\partial\lambda_i}{\partial t}(\lambda_t^{-1}(F(\varphi(x,t),t)),t)$$

for x and t near 0.

Now we substitute this in equation (f):

$$(f'): \quad \frac{\partial}{\partial t}(\lambda_t^{-1}\circ F_t\circ\varphi_t)(x) = \sum_{i=1}^{p} \Big(\frac{\partial\lambda_t^{-1}}{\partial y_i}(F(\varphi(x,t),t)\Big[\frac{\partial F_i}{\partial t}(\varphi(x,t),t)$$

$$-\frac{\partial\lambda_i}{\partial t}(\lambda_t^{-1}(F(\varphi(x,t),t)),t) + \sum_{j=1}^{n}\frac{\partial F_i}{\partial x_j}(\varphi(x,t),t)\frac{\partial\varphi_j}{\partial t}(x,t)\Big]\Big) .$$

Now we claim that the expression in square brackets is identically 0 for each i, $1 \leq i \leq p$, and for x and t near 0. If we can show this we are done, because this implies $\frac{\partial}{\partial t}(\lambda_t^{-1}\circ F_t\circ\varphi_t) = 0$ and hence equation (e) holds.

Now since equation (a) holds everywhere near 0 and since $\varphi(0) = 0$, it also holds near 0 if we replace x by $\varphi(x,t)$. If we also transpose the right-hand side to the left, we then get:

$$(a'): \quad \frac{\partial F_i}{\partial t}(\varphi(x,t),t) - \eta_i(F(\varphi(x,t),t),t)$$

$$+\sum_{j=1}^{n}\frac{\partial F_i}{\partial x_j}(\varphi(x,t),t)\cdot(-\xi_j(\varphi(x,t),t)) = 0 \quad .$$

Now by (d), $-\xi_j(\varphi(x,t),t) = \frac{\partial\varphi_j}{\partial t}(x,t)$ and

$$\eta_i(F(\varphi(x,t),t) = \eta_i(\lambda(\lambda_t^{-1}(F(\varphi(x,t),t)),t),t)$$
$$= \frac{\partial\lambda_i}{\partial t}(\lambda_t^{-1}(F(\varphi(x,t),t)),t) .$$

If we substitute into equation (a'), the left-hand side becomes just the expression in brackets in (f'); the right-hand side is 0, so we are done.

As a special case of this lemma, we have:

Corollary 1.28. Let $F \in m(n+1)$. Suppose there is a germ $\xi \in \mathcal{E}(n+1,n)$ so that for any $x = (x_1,\ldots,x_n)$ near $0 \in R^n$ and any t near $0 \in R$ the following equation holds:

(a) $$\frac{\partial F}{\partial t}(x,t) = \sum_{j=1}^{n}\frac{\partial F_i}{\partial x_j}(x,t)\xi_j(x,t) .$$

Then there is a germ $\varphi \in \mathcal{E}(n+1,n)$ such that for x near 0 in R^n

(b) $$\varphi(x,0) = x$$

and such that for x near 0 in R^n and t near 0 in R

(c) $$F(\varphi(x,t),t) = F(x,0) .$$

φ can be chosen uniquely so that

(d) $$\frac{\partial\varphi_j}{\partial t}(x,t) = -\xi_j(\varphi(x,t),t) \quad \text{for } x \text{ and } t \text{ near } 0.$$

If we define φ_t as before, then for t sufficiently small, φ_t is the germ of a local diffeomorphism of R^n.

Proof: Apply Lemma 1.27 with $p = 1$ and $\eta = 0$.
Then by 1.27(d), $\frac{\partial\lambda}{\partial t} = 0$, so for all t near 0, $\lambda_t = \lambda_o = id_R$.
Then 1.28 (b), (c) and (d) follow from 1.27 (b), (c) and (d).

As another corollary of Lemma 1.27, we have:

<u>Lemma 1.29.</u> Let $F \in \mathcal{E}(n+1)$ and let $F(0) = 0$. Suppose there are germs $\xi \in \mathcal{E}(n+1,n)$ and $\eta \in \mathcal{E}(n+2)$ such that for x near 0 in $\mathbb{R}^n$ and for t near 0 in $\mathbb{R}$ the following equation holds:

(a) $$\frac{\partial F(x,t)}{\partial t} = \sum_{j=1}^{n} \frac{\partial F(x,t)}{\partial x_j} \xi_j(x,t) + \eta(F(x,t),x,t)$$

Then there exist germs $\varphi \in \mathcal{E}(n+1,n)$ and $\lambda \in \mathcal{E}(n+2)$ such that for x near 0 in $\mathbb{R}^n$ and s near 0 in $\mathbb{R}$

(b) $$\varphi(x,0) = x \quad \text{and} \quad \lambda(s,x,0) = s$$

and such that for x near 0 in $\mathbb{R}^n$ and t near 0 in $\mathbb{R}$ we have

(c) $$F(\varphi(x,t),t) = \lambda(F(x,0),x,t)$$

Moreover we may choose φ and λ (in fact uniquely) such that

(d) $$\frac{\partial \varphi_j(x,t)}{\partial t} = -\xi_j(\varphi(x,t),t) \quad (j = 1,\dots,n)$$

$$\text{and } \frac{\partial \lambda}{\partial t}(s,x,t) = \eta(\lambda(s,x,t), \varphi(x,t),t)$$

for $t \in \mathbb{R}$, $x \in \mathbb{R}^n$, $s \in \mathbb{R}$.

<u>Proof:</u> Define $F' \in \mathcal{E}(n+1,n+1)$ by setting $F'(x,t) = (F(x,t),x)$ for $x \in \mathbb{R}^n$, $t \in \mathbb{R}$. Define $\mu \in \mathcal{E}(n+2, n+1)$ by setting $\mu(s,x,t) = (\eta(s,x,t), -\xi_1(x,t),\dots,-\xi_n(x,t))$ for $s \in \mathbb{R}$, $x \in \mathbb{R}^n$, $t \in \mathbb{R}$.

Then we have

(e) $$\frac{\partial F'_i}{\partial t}(x,t) = \sum_{j=1}^{n} \frac{\partial F'_i(x,t)}{\partial x_j} \xi_j(x,t) + \mu_i(F'(x,t),t)$$

for i = 1,...,n+1. When i = 1 this is just 1.29 (a). When i > 1 the left-hand side is clearly 0; the right-hand side reduces to $\xi_{i-1}(x,t)-\xi_{i-1}(x,t) = 0$. Hence (e) holds.

But (e) is just 1.27 (a) with F' for F and μ for η. So by lemma 1.27 there are germs $\varphi \in \mathcal{E}(n+1,n)$ and $\Lambda \in \mathcal{E}(n+2,n+1)$ with

(f) $\varphi(x,0) = x$ and $\Lambda(s,x,0) = (s,x)$ for $x \in R^n$, $s \in R$

(g) $F'(\varphi(x,t),t) = \Lambda(F'(x,0),t)$ for $x \in R^n$, $t \in R$

and (h) $\dfrac{\partial\varphi_j(x,t)}{\partial t} = -\xi_j(\varphi(x,t),t)$ $(j = 1,\ldots,n)$

$$\frac{\partial\Lambda_i}{\partial t}(s,x,t) = \mu_i(\Lambda(s,x,t),t) \quad (i = 1,\ldots,n+1)$$

for $x \in R^n$, $s \in R$, $t \in R$.

Now let $\lambda = \Lambda_1$. Since $F = F'_1$, equations (b) and (c) clearly follow from (f) and (g) above. We must still prove (d). Now we claim that if i > 1, then $\Lambda_i(s,x,t) = \varphi_{i-1}(x,t)$. For by (h), the germs $\Lambda_2,\ldots,\Lambda_{n+1}$ satisfy the differential equations:

$$\frac{\partial\Lambda_{j+1}}{\partial t}(s,x,t) = \mu_{j+1}(\Lambda(s,x,t),t)$$

$$= -\xi_j(\Lambda_2(s,x,t),\ldots,\Lambda_{n+1}(s,x,t),t)$$

$$(j = 1,\ldots,n).$$

But by (h), if we replace $\Lambda_{j+1}(s,x,t)$ by $\varphi_j(x,t)$, the same set of differential equations holds; moreover

$\Lambda_{j+1}(s,x,0) = x_j = \varphi_j(x,0)$ $(x \in R^n,\ s \in R,\ 1 \leq j \leq n)$.

Since the solution of this system of differential equations is uniquely determined by the initial conditions, we have $\Lambda_{j+1}(s,x,t) = \varphi_j(x,t)$ for $s \in \mathbb{R}$, $x \in \mathbb{R}^n$, $t \in \mathbb{R}$, $j = 1,\dots,n$. But then, again by (h), we find that

$$\frac{\partial\lambda}{\partial t}(s,x,t) = \frac{\partial\Lambda_1}{\partial t}(s,x,t) = \mu_1(\Lambda(s,x,t),t) = \eta(\lambda(s,x,t),\varphi(x,t),t).$$

Hence clearly(d) holds. And φ and λ are in fact the <u>unique</u> solution of the differential equations (d) with initial conditions (b). This completes the proof.

We conclude with a trivial remark on differential equations to which we shall refer in the course of several proofs in later chapters.

<u>Remark 1.30.</u> Let $f \in \mathcal{E}(n+r+1,p)$ and suppose $f(0) = 0$. Let $\xi \in \mathcal{E}(p+r+1,p)$. Suppose f satisfies the differential equations

$$\frac{\partial f_i}{\partial t}(x,y,t) = \xi_i(f(x,y,t),y,t) \quad (i = 1,\dots,p;\ x \in \mathbb{R}^n,\ y \in \mathbb{R}^r,\ t \in \mathbb{R})$$

and suppose $f(x,y,0)$ depends only on $y \in \mathbb{R}^r$ and <u>not</u> on $x \in \mathbb{R}^n$. Then $f(x,y,t)$ depends only on y and t, but not on x.

<u>Proof:</u> Choose representatives for f and ξ defined near the origin of $\mathbb{R}^{n+r+1}$ and $\mathbb{R}^{p+r+1}$ respectively; for convenience we use the same names f and ξ for the representatives.

For x near $0 \in \mathbb{R}^n$ define $f_x \in \mathcal{E}(r+1,p)$ by $f_x(y,t) = f(x,y,t)$ $(y \in \mathbb{R}^r,\ t \in \mathbb{R})$.

Now for each x near 0 in $\mathbb{R}^n$, the germ f_x satisfies the differential equations

$$\frac{\partial (f_x)_i}{\partial t}(y,t) = \xi_i(f_x(y,t),y,t) \quad (i = 1,\ldots,p)$$

and with the same initial conditions for every x (since f(x,y,0) is independent of x). Hence since these equations have solutions which are uniquely determined by the initial conditions, f(x,y,t) is independent of x.

§ 2 FINITELY DETERMINED GERMS

For convenience we shall agree that throughout this chapter, unless otherwise stated, x shall denote an element of $\mathbb{R}^n$ and we shall denote the standard coordinates on $\mathbb{R}^n$ by $x_1, \ldots, x_n$.

<u>Definition 2.1.</u> Let f and g be germs in $\mathcal{E}(n)$. We say that f and g are <u>right-equivalent</u> (and we write $f \sim_r g$) if there is a $\varphi \in L(n)$ such that $f = g\varphi$.

<u>Definition 2.2.</u> Let f and g be germs in $m(n)$. We say f and g are <u>right-left equivalent</u> or <u>two-sided equivalent</u> (and we write $f \sim_{rl} g$) if there is a $\varphi \in L(n)$ and a $\psi \in L(1)$ such that $f = \psi g \varphi$.

Obviously right equivalence and two-sided equivalence are equivalence relations. Note that for germs f and g in $\mathcal{E}(n)$ to be right-equivalent, it is necessary that $f(0) = g(0)$. Moreover if $f(0) = g(0)$ and if we define f' and g' in $m(n)$ by $f'(x) := f(x) - f(0)$ and $g'(x) := g(x) - g(0)$, then $f \sim_r g$ if and only if $f' \sim_r g'$. So for simplicity we may henceforth without any loss of generality restrict our attention to germs in $m(n)$.

<u>Definition 2.3.</u> Let $f \in m(n)$, and let k be a non-negative integer. We say f is <u>right k-determined</u> (abbreviated <u>r k-determined</u>) if for any $g \in m(n)$ such that $j^k g(0) = j^k f(0)$ it is true that $f \sim_r g$. Similarly we say f is <u>right-left k-determined</u>, or <u>rl k-determined</u>, if for any $g \in m(n)$ such that $j^k g(0) = j^k f(0)$

it is true that $f \sim_{rl} g$.

If f is right (or right-left) k-determined for some integer k, then we say f is right (or right-left) finitely determined. Note that unless $n = 0$ no germ in $m(n)$ is 0-determined.

Clearly the property of being k-determined depends only on the k-jet (at 0) of the germ. So we may make the following definition.

Definition 2.4. Let $z \in J_0^k(n,1)$. Then z is said to be right (right-left) k-determining if for some f in $m(n)$ such that $j^k f(0) = z$ (and hence for any such f) it is true that f is right (right-left) k-determined.

We make the additional observation that the property of being right (or right-left) k-determined is invariant under right (respectively right-left) equivalence. Suppose, for example, that $f \in m(n)$ is r k-determined and suppose $f \sim_r g$. Then there is a $\varphi \in L(n)$ such that $f = g\varphi$. Now if $h \in m(n)$ and if $j^k h(0) = j^k g(0)$, then $j^k(h\varphi)(0) = j^k(g\varphi)(0) = j^k f(0)$ (by the observation preceding Definition 1.25). Hence $h \sim_r h\varphi \sim_r f \sim_r g$. So g is r k-determined. A similar argument works for the two-sided case.

Finally we observe that if $f \in m(n)$ is non-singular (i.e. $f \notin m(n)^2$) then f is right 1-determined, for any germ with the same 1-jet is also non-singular, and any non-singular germ is right-equivalent for example to the germ x_1 of the first coordinate function, so all non-singular germs are right equivalent to each other.

We may apply Definition 2.3 to jets as well as germs:

Definition 2.5. Let $z \in J_o^q(n,1)$, and let $k \leq q$. Let

$$A = \{z' \in J_o^q(n,1) \mid \pi_{q,k}(z') = \pi_{q,k}(z)\} .$$

Then z is said to be right k-determined if $A \subseteq zL^q(n)$ and z is said to be right-left k-determined if $A \subseteq L^q(1)zL^q(n)$.

Obviously if $f \in m(n)$ and f is right (right left) k-determined, and if $q \geq k$, then $\pi_q f \in J_o^q(n,1)$ is right (right-left) k-determined. We shall see (Corollary 2.11) that if q is big enough, then the converse holds too. Of course, in general k-determinacy of jets is weaker than k-determinacy of germs which represent them; for example, any k-jet is trivially k-determined in both senses.

We now give some results which relate k-determinacy to certain algebraic conditions. Our first proposition is due to Mather ([6, Theorem 3.5 for the group $\mathcal{L}$] or, in the form in which we give it here, [10, Proposition 1]).

Theorem 2.6. (Mather). Let $f \in m(n)$. Let k be a non-negative integer. Suppose that (as a condition on ideals of $\mathcal{E}(n)$):

(a) $$m(n)^k \subseteq m(n)\langle \partial f/\partial x_1, \dots, \partial f/\partial x_n \rangle + m(n)^{k+1} .$$

Then f is right-k-determined.

Proof: Let $g \in m(n)$ and suppose $j^k f(0) = j^k g(0)$. If $t \in \mathbb{R}$, define $F_t \in m(n)$ by $F_t(x) := (1-t)f(x) + tg(x)$. Then $f = F_0$ and $g = F_1$.

So to prove the theorem, we must show that $F_0 \sim_r F_1$. We shall do this by showing, in fact, that for any $t \in [0,1] \subset \mathbb{R}$, if t' is sufficiently close to t then $F_{t'} \sim_r F_t$. This will obviously imply $F_0 \sim_r F_1$, so we shall be done.

So let $t \in [0,1]$ be given. Define a germ $H \in \mathcal{E}(n+1)$ by $H(x,s) = F_{s+t}(x)$ for x near 0 in $\mathbb{R}^n$ and s near 0 in $\mathbb{R}$.

Now suppose we can find a germ $\xi \in \mathcal{E}(n+1,n)$ such that

(b) $\quad \frac{\partial H}{\partial s}(x,s) = \sum_{j=1}^{n} \frac{\partial H}{\partial x_j}(x,s)\xi_j(x,s)$, for $x \in \mathbb{R}^n$ and $s \in \mathbb{R}$

and such that

(c) $\quad \xi(0,s) = 0$ for s in $\mathbb{R}$.

Then because of (b) we may apply corollary 1.28 to find a germ $\varphi \in \mathcal{E}(n+1,n)$ such that

(d) $\quad \varphi(x,0) = x$

(e) $\quad H(\varphi(x,s),s) = H(x,0)$ and

(f) $\quad \frac{\partial \varphi_j}{\partial s}(x,s) = -\xi_j(\varphi(x,s),s)$

for x near 0 in $\mathbb{R}^n$ and s near 0 in $\mathbb{R}$.

First we observe that $\varphi(0,s) = 0$ for s near 0 in $\mathbb{R}$. This is so because $\varphi(0,s)$, as a function of s, is the unique solution of the differential equation $\frac{\partial \varphi_j}{\partial s}(0,s) = -\xi_j(\varphi(0,s),s)$ with initial condition $\varphi(0,0) = 0$. But because of (c), the constant function 0 is also a solution, so $\varphi(0,s) \equiv 0$.

Now if, for s near 0 in $\mathbb{R}$, we define $\varphi_s \in \mathcal{E}(n,n)$ by $\varphi_s(x) = \varphi(x,s)$, then $\varphi_s(0) = 0$ for s small, and by (d) φ_s

is a germ of a local diffeomorphism for s small, so if s is near 0 then $\varphi_s \in L(n)$. And from (e) it follows that if s is near 0, then $F_t = F_{s+t} \circ \varphi_s$ and hence $F_t \sim_r F_{s+t}$, which is just what we wanted to show.

So what remains to be proved is that there is a $\xi \in \mathcal{E}(n+1,n)$ satisfying (b) and (c).

Now let $\pi\colon R^{n+1} \longrightarrow R^n$ be the canonical projection, defined by $\pi(x_1,\ldots,x_n,s) = (x_1,\ldots,x_n)$, and consider $\mathcal{E}(n)$ as a subring of $\mathcal{E}(n+1)$ via the injective ring-homomorphism π^*. Then condition (a) implies

(g) $$m(n)^k\mathcal{E}(n+1) \subseteq m(n)\langle \partial f/\partial x_1,\ldots,\partial f/\partial x_n\rangle_{\mathcal{E}(n+1)} + m(n)^{k+1}\mathcal{E}(n+1)\ .$$

Now we observe that for any i, $1 \leq i \leq n$, we have
$\frac{\partial H}{\partial x_i}(x,s) - \frac{\partial f}{\partial x_i}(x) = (s+t)\left(\frac{\partial g}{\partial x_i}(x) - \frac{\partial f}{\partial x_i}(x)\right)$ and since $g-f \in m(n)^{k+1}$, it follows that $\partial H/\partial x_i - \partial f/\partial x_i \in m(n)^k\mathcal{E}(n+1)$. Hence (g) remains valid if we replace $\partial f/\partial x_i$ by $\partial H/\partial x_i$. So we have

(h) $$m(n)^k\mathcal{E}(n+1) \subseteq \langle \partial H/\partial x_1,\ldots,\partial H/\partial x_n\rangle_{m(n)\mathcal{E}(n+1)} + m(n)^{k+1}\mathcal{E}(n+1)\ .$$

Now we apply Nakayama's lemma (Lemma 1.13) (with $I = m(n)\mathcal{E}(n+1)$) and deduce

(j) $$m(n)^k\mathcal{E}(n+1) \subseteq \langle \partial H/\partial x_1,\ldots,\partial H/\partial x_n\rangle_{m(n)\mathcal{E}(n+1)}\ .$$

Clearly $\frac{\partial H}{\partial s} \in m(n)^k\mathcal{E}(n+1)$, so we can find $\xi_1,\ldots,\xi_n \in m(n)\mathcal{E}(n+1)$ so that

$$\frac{\partial H}{\partial s} = \sum_{i=1}^{n} \frac{\partial H}{\partial x_i}\,\xi_i\ .$$

These ξ_i give us a $\xi \in \mathcal{E}(n+1,n)$ satisfying (b), and since each $\xi_i \in m(n)\mathcal{E}(n+1)$, (c) also holds, so we are done.

<u>Corollary 2.7.</u> Let $f \in m(n)$. Let k be a non-negative integer. If

(a) $m(n)^k \subseteq \langle \partial f/\partial x_1, \ldots, \partial f/\partial x_n \rangle_{m(n)} + f^* m(1) + m(n)^{k+2}$

then f is right k+1-determined.

<u>Proof:</u> Without loss of generality we may assume $f \in m(n)^2$, since otherwise f is non-singular and hence right 1-determined. Now apply Theorem 1.19 (with $A = f^* m(1)$, $B = \langle \partial f/\partial x_i \rangle_{m(n)}$, $C = \mathcal{E}(n)$ and $D = m(n)^k$). Then by 1.19 (c) we have

$m(n)^{k+1} \subseteq \langle \partial f/\partial x_1, \ldots, \partial f/\partial x_n \rangle_{m(n)}$ and hence, by Theorem 2.6, f is right k+1-determined.

This corollary is in fact the analog, for right-left determinacy, of theorem 2.6, in spite of the fact that the conclusion is that f is right k+1-determined (obviously f is then also rl k+1-determined). We do not know whether, under the conditions of corollary 2.7, f is in fact rl k-determined.

Mather also has proved an analog of 2.6 for the "two-sided" case (see [6, Theorem 3.5 for the group $\mathcal{A}$, proof in § 6]). However, though his hypotheses are more general (he considers germs in $\mathcal{E}(n,p)$ for any p), his conclusions are much weaker than ours (under similar hypotheses to ours, he proves only that f is right-left $(k + \binom{n+k-1}{n} + 2)$-determined).

The last two results give sufficient conditions for a germ to be k-determined; we now give necessary conditions for k-determinacy, which are unfortunately weaker. We begin with a lemma from which these conditions follow. This lemma is due to Mather [6, § 7].

<u>Lemma 2.8.</u> Let $f \in m(n)$, let k be a non-negative integer, and let $z := \pi_k(f) \in J_o^k(n,1)$.

Then

(a) $\pi_k^{-1} T_z z L^k(n) = m(n)\langle \partial f/\partial x_1, \ldots, \partial f/\partial x_n \rangle + m(n)^{k+1}$

and

(b) $\pi_k^{-1} T_z L^k(1) z L^k(n) = m(n)\langle \partial f/\partial x_1, \ldots, \partial f/\partial x_n \rangle + f^* m(1) + m(n)^{k+1}$

(We recall that $\pi_k: \mathcal{E}(n) \longrightarrow J^k(n,1)$ is the canonical projection and note that $J_o^k(n,1) = \pi_k(m(n))$. See definitions 1.20 and 1.21)

<u>Proof:</u> Suppose in general that G is any Lie group acting on a manifold M, and let 1 be the identity of G and x be any point of M. Then we can define a map $\alpha: G \longrightarrow M$ by $\alpha(g) = gx$; α induces a linear map $T\alpha$ from the tangent space of G at 1 into the tangent space of M at x; the image of $T\alpha$ is then just the tangent space at x of the orbit of x.

Moreover suppose v is a tangent vector to G at 1, and suppose $w: \mathbb{R} \longrightarrow G$ is a smooth path with $w(0) = 1$ and such that v is the velocity vector of w at 0. Then $T\alpha(v)$ is just the velocity vector at 0 of the path $\alpha w: \mathbb{R} \longrightarrow M$. We apply these observations in what follows.

$L^k(n)$ is an open subset of $J_o^k(n,n)$, which is canonically isomorphic to a Euclidean space, so we may identify the tangent plane to $L^k(n)$ at the identity with $J_o^k(n,n)$ itself. Similarly the tangent plane to $L^k(1)$ at the identity may be identified with $J_o^k(1,1)$, so the tangent plane at the identity to $L^k(1) \times L^k(n)$ is just $J_o^k(1,1) \times J_o^k(n,n)$. If we identify $L^k(n)$

with the subgroup $Id_{\mathbb{R}} \times L^k(n)$ of $L^k(1) \times L^k(n)$ we have a corresponding inclusion of tangent planes.

Let $(\lambda,\beta) \in J_0^k(1,1) \times J_0^k(n,n)$ be a tangent vector to $L^k(1) \times L^k(n)$ at the identity. If $\lambda = 0$ then (λ,β) is tangent to $L^k(n)$. Now let $\lambda' \in m(1)$ represent λ and let $\beta' \in \mathcal{E}(n,n)$ represent β. (If $\lambda = 0$ choose $\lambda' = 0$, for simplicity). For $t \in \mathbb{R}$ define $\eta_t \in m(1)$ by $\eta_t = Id_{\mathbb{R}}+t\lambda'$ and define $\delta_t \in \mathcal{E}(n,n)$ by $\delta_t = Id_{\mathbb{R}^n} + t\beta'$. Then (λ,β) is just the velocity vector at 0 of the path $t \longrightarrow (\pi_k\eta_t, \pi_k\delta_t)$. So by our observations above, an element of $T_z L^k(1) z L^k(n)$ has the form

$$\frac{\partial}{\partial t}(\pi_k\eta_t \cdot z \cdot \pi_k\delta_t)\Big|_{t=0} = \frac{\partial}{\partial t}\,\pi_k(\eta_t \cdot f \cdot \delta_t)\Big|_{t=0} = \pi_k\Big(\frac{\partial}{\partial t}(\eta_t \cdot f \cdot \delta_t)\Big|_{t=0}\Big)$$

$$= \pi_k\Big(\frac{\partial}{\partial t}\Big(f(Id_{\mathbb{R}^n}+t\beta')+t\lambda'(f(Id_{\mathbb{R}^n}+t\beta'))\Big)\Big|_{t=0}\Big)$$

$$= \pi_k\Big(\Sigma_{i=1}^n \frac{\partial f}{\partial x_i} \cdot \beta_i' + \lambda' \cdot f\Big).$$

So since the only conditions on β' and λ' were that $\beta'(0) = 0$ and $\lambda'(0) = 0$ we see that

$$T_z L^k(1) z L^k(n) = \pi_k(\langle \partial f/\partial x_1, \ldots, \partial f/\partial x_n \rangle_{m(n)} + f^* m(1)),$$

which proves (b).

If $\lambda = 0$ we chose $\lambda' = 0$ so the same computation shows that $T_z z L^k(n) = \pi_k(m(n)\langle \partial f/\partial x_1, \ldots, \partial f/\partial x_n \rangle)$ which proves (a).

<u>Corollary 2.9.</u> Let q be a non-negative integer. Let $z \in J_o^q(n,1)$ and let $f \in m(n)$ be a representative of z. Let $k \leq q$. If z is right k-determined, then

(a) $$m(n)^{k+1} \subseteq m(n)\langle \partial f/\partial x_1,\dots,\partial f/\partial x_n\rangle + m(n)^{q+1} .$$

If z is right-left k-determined, then

(b) $$m(n)^{k+1} \subseteq m(n)\langle \partial f/\partial x_1,\dots,\partial f/\partial x_n\rangle + f^*m(1) + m(n)^{q+1} .$$

<u>Proof:</u> Let $A = \{z' \in J_o^q(n,1) \mid \pi_{q,k}z' = \pi_{q,k}z\}$

A is an affine subspace of $J_o^q(n,1)$ and $T_zA = \pi_q(m(n)^{k+1})$.

Now if z is r k-determined, then $A \subseteq zL^q(n)$ so $T_zA \subseteq T_z zL^q(n)$. Then (a) follows immediately from 2.8 (a). Similarly if z is rl k-determined, then $T_zA \subseteq T_z L^q(1)zL^q(n)$, so (b) follows from 2.8 (b).

<u>Corollary 2.10.</u> Let $f \in m(n)$. Let k be a non-negative integer.

(a) If f is right k-determined then

$$m(n)^{k+1} \subseteq m(n)\langle \partial f/\partial x_1,\dots,\partial f/\partial x_n\rangle$$

(b) If f is right-left k-determined then

$$m(n)^{k+1} \subseteq m(n)\langle \partial f/\partial x_1,\dots,\partial f/\partial x_n\rangle + f^*m(1) .$$

<u>Proof:</u> (a) If f is r k-determined, then so is $\pi_{k+1}(f)$. The result then follows from 2.9 (a) and Lemma 1.13 (Nakayama).

(b) If f is rl k-determined then so is $\pi_{k+2}(f)$. The result then follows from 2.9 (b) and Theorem 1.19 (Malgrange).

Corollary 2.11. Let $f \in m(n)$. Let k be a non-negative integer.

(a) If $\pi_{k+1}(f)$ is right k-determined then f is right k-determined.

(b) If $\pi_{k+2}(f)$ is right-left k-determined then f is right-left k-determined.

Proof: (a) Let $g \in m(n)$ and suppose $\pi_k(g) = \pi_k(f)$. Since $\pi_{k+1}(f)$ is r k-determined, there is a $\varphi \in L(n)$ such that $\pi_{k+1}(f) = \pi_{k+1}(g\varphi)$, and also, by Corollary 2.9, equation 2.9 (a) holds with $q = k+1$. But then, by theorem 2.6, f is r k+1-determined so $f \sim_r g\varphi \sim_r g$.

(b) If $\pi_{k+2}(f)$ is right-left k-determined then equation 2.9 (b) holds with $q = k+2$, and hence by corollary 2.7 f is right k+2-determined. But then it follows that f is right-left k-determined, for if $g \in m(n)$ and $\pi_k(g) = \pi_k(f)$, then since $\pi_{k+2}(f)$ is rl k-determined there are a $\varphi \in L(n)$ and a $\psi \in L(1)$ such that $\pi_{k+2}(f) = \pi_{k+2}(\psi g\varphi)$ and since f is right k+2-determined we have $f \sim_r \psi g\varphi \sim_{rl} g$, so $f \sim_{rl} g$.

The proofs of 2.9 (a), 2.10 (a), and 2.11 (a) are given by Mather in [10,Propositions 2 and 3]. Also, in [6, § 8] he proves a weaker and less explicit form of 2.10 (b).

Corollary 2.12. Let $f \in m(n)$ and let k be finite. If f is right-left k-determined, then f is right k+2-determined.

Proof: If f is rl k-determined, then clearly $\pi_{k+2}(f)$ is rl k-determined. The proof of 2.11 (b) then shows f is right k+2-determined.

Corollary 2.13. Let $f \in m(n)$. Then f is right finitely determined if and only if f is right-left finitely determined.

Proof: Immediate from Corollary 2.12.

So for the concept finitely determined (but not for k-determined!) we may in future omit the modifiers right and right-left.

Definition 2.14. Let $f \in m(n)$. We shall write $\langle \partial f/\partial x \rangle$ as an abbreviation for $\langle \partial f/\partial x_1, \ldots, \partial f/\partial x_n \rangle$.
We define:

$$\tau(f) = \dim_R \mathcal{E}(n)/\langle \partial f/\partial x \rangle$$

and $$\sigma(f) = \dim_R \mathcal{E}(n)/(\langle \partial f/\partial x \rangle + f^*\mathcal{E}(1)) .$$

If k is a non-negative integer, we define

$$\tau_k(f) = \dim_R \mathcal{E}(n)/(\langle \partial f/\partial x \rangle + m(n)^k)$$

and $$\sigma_k(f) = \dim_R \mathcal{E}(n)/(\langle \partial f/\partial x \rangle + f^*\mathcal{E}(1)+m(n)^k).$$

Clearly, if $k \leq q$ then $\tau_k(f) \leq \tau_q(f) \leq \tau(f)$ and $\sigma_k(f) \leq \sigma_q(f) \leq \sigma(f)$.

We now wish to investigate the relationship between $\tau_k(f)$, $\tau(f)$, $\sigma_k(f)$, $\sigma(f)$ and finite determinacy of f. Before stating our result, we give a lemma which will be needed in the proof.

Lemma 2.15. Let $f \in m(n)^2$.
Then $f \in m(n)\langle \partial f/\partial x_1, \ldots, \partial f/\partial x_n \rangle + m(n)^3$.

<u>Proof:</u> Let $g = f - \frac{1}{2}\Sigma_{i=1}^{n} x_i \partial f/\partial x_i$. Since $f \in m(n)^2$, an elementary computation shows $g \in m(n)^3$. Clearly this proves the lemma.

<u>Corollary 2.16.</u> Let $f \in m(n)$. Let k be a non-negative integer.

(a) If $\tau_k(f) < k$ then $m(n)^k \subseteq m(n)\langle \partial f/\partial x_1, \ldots, \partial f/\partial x_n\rangle$ and f is right k-determined.

(b) If $\sigma_{3k-1}(f) < 2k-1$ then $m(n)^{3k-1} \subseteq m(n)\langle \partial f/\partial x_1, \ldots, \partial f/\partial x_n\rangle$ and f is right 3k-1-determined.
If $\sigma_{3k}(f) < 2k$ then $m(n)^{3k} \subseteq m(n)\langle \partial f/\partial x_1, \ldots, \partial f/\partial x_n\rangle$ and f is right 3k-determined.

<u>Proof:</u> We write $\langle \partial f/\partial x\rangle$ as an abbreviation for $\langle \partial f/\partial x_1, \ldots, \partial f/\partial x_n\rangle$.

(a) If $\tau_k(f) < k$ then $k > 0$ and by Corollary 1.14 we have

$$m(n)^{k-1} \subseteq \langle \partial f/\partial x\rangle.$$

Hence $m(n)^k \subseteq m(n)\langle \partial f/\partial x\rangle$ and by theorem 2.6 f is right k-determined.

(b) Let q be either 3k or 3k-1 and let $\sigma_q(f) = r$. Then

$$r = \dim_{\mathbb{R}} \mathcal{E}(n)/(\langle \partial f/\partial x\rangle + f^*\mathcal{E}(1) + m(n)^q).$$

Now we may suppose $f \in m(n)^2$, since otherwise f is non-singular and hence right 1-determined. But if $f \in m(n)^2$, then by Lemma 2.15, $f \in \langle \partial f/\partial x\rangle + m(n)^3$ and hence $f^k \in \langle \partial f/\partial x\rangle + m(n)^{3k}$ $\subseteq \langle \partial f/\partial x\rangle + m(n)^q$. Hence $\langle \partial f/\partial x\rangle + f^*\mathcal{E}(1) + m(n)^q$ $= \langle \partial f/\partial x\rangle + \langle 1, f, f^2, \ldots, f^{k-1}\rangle_{\mathbb{R}} + m(n)^q$.

So clearly $\tau_q(f) = \dim_{\mathbb{R}} \mathcal{E}(n)/(\langle \partial f/\partial x\rangle + m(n)^q) \leq r+k$.

Consequently, if $q = 3k-1$ and if $r < 2k-1$, then $\tau_{3k-1}(f) < 3k-1$. And if $q = 3k$ and if $r < 2k$ then $\tau_{3k}(f) < 3k$.

In either case the conclusion then follows by part (a).

Part (a) is due to Mather [6, Theorems 3.5, 3.6 for the group $\mathcal{R}$; and 10, p.21]. (Proposition 3.6 of [6] also contains a much weaker form of part (b)).

We may consolidate our results thus far to give the following characterization of finite determinacy:

<u>Corollary 2.17.</u> Let $f \in m(n)$. The following statements are equivalent:

(a) f is finitely determined

(b) For some finite k

$$m(n)^k \subseteq \langle \partial f/\partial x_1, \ldots, \partial f/\partial x_n \rangle$$

(c) For some finite k

$$m(n)^k \subseteq \langle \partial f/\partial x_1, \ldots, \partial f/\partial x_n \rangle + f^* \mathcal{E}(1)$$

(d) For some finite k

$$\tau_k(f) < k$$

(e) For some finite k

$$\sigma_{3k}(f) < 2k$$

(f) $\tau(f) < \infty$

(g) $\sigma(f) < \infty$

<u>Proof:</u> (a) $\Rightarrow$ (b) by 2.10; (b) $\Rightarrow$ (f) $\Rightarrow$ (d) is trivial; (d) $\Rightarrow$ (a) by 2.16. Also, (b) $\Rightarrow$ (c) $\Rightarrow$ (g) $\Rightarrow$ (e) is trivial; (e) $\Rightarrow$ (a) by 2.16.

This result was in part first proved by Tougeron [17] .

Definition 2.18. Let $f \in m(n)^2$. Then the right codimension of f is:

$$\text{r-codim}(f) = \dim_{\mathbb{R}} m(n)/\langle \partial f/\partial x_1, \ldots, \partial f/\partial x_n \rangle .$$

The right-left codimension of f is

$$\text{rl-codim}(f) = \dim_{\mathbb{R}} m(n)/(\langle \partial f/\partial x_1, \ldots, \partial f/\partial x_n \rangle + f^* m(1)) .$$

One easily sees that r-codim(f) and rl-codim(f) depend only on the rl-equivalence class of f. For if $\varphi \in L(n)$, $\psi \in L(1)$ and $g = \psi f \varphi$, then observe that φ^* is an $\mathbb{R}$-algebra automorphism of $m(n)$, and $\langle \partial g/\partial x_1, \ldots, \partial g/\partial x_n \rangle = \varphi^*(\langle \partial f/\partial x_1, \ldots, \partial f/\partial x_n \rangle)$ (as one sees by computation) and $g^* m(1) = \varphi^* f^* \psi^* m(1) = \varphi^*(f^* m(1))$.

Clearly, $\text{r-codim}(f) = \tau(f) - 1$ and $\text{rl-codim}(f) = \sigma(f)$. So if f is finitely determined, its right- and right-left codimensions are finite, and if either the right- or the right-left codimension of f is finite, then f is finitely determined. In fact one easily sees by Corollary 2.16 that if $\text{r-codim}(f) = k < \infty$, then f is right k+2-determined, and if rl-codim $f = q < \infty$, then if q is even, f is right $\frac{3}{2}q+2$-determined, and if q is odd, f is right $\frac{3}{2}(q+1)$-determined.

The following proposition justifies the use of the term "codimension".

Proposition 2.19. Let $f \in m(n)^2$ be a finitely determined singularity. Let k be a non-negative integer and let $z = \pi_k(f) \in J_0^k(n,1)$. Let $s = \text{r-codim}(f)$ and let $q = \text{rl-codim}(f)$.

If either $k > s$ or $k > \frac{3}{2} q$, then the codimension of $zL^k(n)$ in $J_o^k(n,1)$ is $n+s$, and the codimension of $L^k(1)zL^k(n)$ in $J_o^k(n,1)$ is $n+q$.

Proof: We shall write $\langle \partial f/\partial x \rangle$ as an abbreviation for $\langle \partial f/\partial x_1, \ldots, \partial f/\partial x_n \rangle$.

We point out that if $k > \frac{3}{2} q$, then if q is even we have $k \geq \frac{3}{2} q+1$ and if q is odd we have $k \geq \frac{3}{2} q+\frac{1}{2}$, since k is an integer.

So under either of the hypotheses on k (namely $k > s$ or $k > \frac{3}{2} q$) we have

$$m(n)^{k+1} \subseteq m(n)\langle \partial f/\partial x \rangle$$

by corollary 2.16 (see the remarks preceding this proposition).

Now by lemma 2.8, and since $m(n)^{k+1} \subseteq m(n)\langle \partial f/\partial x \rangle$, we have

$$(*) \quad \begin{aligned} \operatorname{codim} zL^k(n) &= \dim_R J_o^k(n,1)/\pi_k(m(n)\langle \partial f/\partial x \rangle) \\ &= \dim_R m(n)/m(n)\langle \partial f/\partial x \rangle \qquad \text{and} \\ \operatorname{codim} L^k(1)zL^k(n) &= \dim_R J_o^k(n,1)/\pi_k(m(n)\langle \partial f/\partial x \rangle + f^* m(1)) \\ &= \dim_R m(n)/(m(n)\langle \partial f/\partial x \rangle + f^* m(1)) \,. \end{aligned}$$

If we can show

$$(**) \quad \begin{aligned} &\dim_R \frac{\langle \partial f/\partial x \rangle}{m(n)\langle \partial f/\partial x \rangle} = n \qquad \text{and} \\ &\dim_R \frac{\langle \partial f/\partial x \rangle + f^* m(1)}{m(n)\langle \partial f/\partial x \rangle + f^* m(1)} = n \end{aligned}$$

we are clearly done, by the equations (*) and by the definitions of r-codim(f) and rl-codim f.

To prove (**) it is enough to show that $\frac{\partial f}{\partial x_1}, \ldots, \frac{\partial f}{\partial x_n}$ are linearly independent over $\mathbb{R}$ modulo $m(n)\langle \partial f/\partial x\rangle + f^* m(1)$. (They are then also linearly independent modulo $m(n)\langle \partial f/\partial x\rangle$).

So suppose we can find real numbers $c_1, \ldots, c_n$, not all 0, and elements $u_1, \ldots, u_n$ of $m(n)$ and an element w of $m(1)$ such that

$$(***) \quad \sum_{i=1}^{n} c_i \partial f/\partial x_i = \sum_{j=1}^{n} u_j \partial f/\partial x_j + f^*(w) .$$

We shall derive a contradiction.

Define a germ X at 0 of a vector field on $\mathbb{R}^n$ by setting

$$X = \sum_{i=1}^{n} (c_i - u_i) \frac{\partial}{\partial x_i} .$$

The c_i are not all equal to zero; without loss of generality we may assume $c_1 \neq 0$.

Now $m(n)^{k+1} \subseteq \langle \partial f/\partial x\rangle$; in particular $x_1^{k+1} \in \langle \partial f/\partial x\rangle$.

Let h be the smallest integer such that $x_1^h \in \langle \partial f/\partial x\rangle$.

Note that $h \geq 1$, since $f \in m(n)^2$ and hence $\langle \partial f/\partial x\rangle \subseteq m(n)$.
We shall derive a contradiction by showing that $x_1^{h-1} \in \langle \partial f/\partial x\rangle$.
To show this it is enough to show that $X(x_1^h) \in \langle \partial f/\partial x\rangle$. For by the definition of X we have $X(x_1^h) = (c_1 - u_1) h x_1^{h-1}$. Now $c_1 \neq 0$ and $h \geq 1$ and $u_1 \in m(n)$; hence $h(c_1 - u_1)$ is a unit of $\mathcal{E}(n)$ and so if $X(x_1^h) \in \langle \partial f/\partial x\rangle$, then also $x_1^{h-1} \in \langle \partial f/\partial x\rangle$.

Now since $x_1^h \in \langle \partial f/\partial x\rangle$ there are germs $\xi_1, \ldots, \xi_n \in \mathcal{E}(n)$ such that

$$x_1^h = \sum_{i=1}^{n} \xi_i \partial f/\partial x_i .$$

Applying X to this equation, we get

$$X(x_1^h) = \sum_{i=1}^{n} X(\xi_i)\frac{\partial f}{\partial x_i} + \sum_{i=1}^{n} \xi_i X(\partial f/\partial x_i) \quad .$$

We claim that for each i, $1 \leq i \leq n$ we have $X(\partial f/\partial x_i) \in \langle \partial f/\partial x \rangle$; this will imply $X(x_1^h) \in \langle \partial f/\partial x \rangle$ so if we can show this we are done.

Now clearly $X(\partial f/\partial x_i) = \frac{\partial}{\partial x_i} X(f)$. But it follows from (***) and the definition of X that $X(f) = f^*(w) = w \circ f$. Hence $\frac{\partial}{\partial x_i} X(f) = \left(\frac{\partial w}{\partial t} \circ f\right)\frac{\partial f}{\partial x_i} \in \langle \partial f/\partial x \rangle$ (here t is the coordinate of $\mathbb{R}$). This completes the proof.

Mather defines the right-codimension in [10] and proves an easier version of Proposition 2.19 in which he considers only the $L^k(n)$ orbit (see [10, Lemma 2]).

We conclude this section by defining, for future reference, some useful algebraic subsets of the jet spaces. The definitions and propositions below are taken from Mather [10, pp.21-22] .

First, an observation. It is easily seen that if k is a non-negative integer and if $f \in m(n)$, then $\tau_k(f)$ depends only on the <u>k-jet</u> of f at 0. So the following definition makes sense.

<u>Definition 2.20.</u> Let k be a non-negative integer.

If $z \in J_0^k(n,1)$, define

$\tau(z) = \tau_k(f)$, where f is any germ in $m(n)$ such that $\pi_k(f) = z$.

By our remark, the definition of $\tau(z)$ does not depend on the choice of f.

Now we define:

Definition 2.21. Let k be a non-negative integer. We set $Z_k := \{z \in J_o^k(n,1) \mid \tau(z) \geq k\}$.

From corollary 2.16 it is clear that if $z \in J_o^k(n,1)$ and if $z \notin Z_k$, then z is right k-determining. Conversely, Corollary 2.17 implies that if $f \in m(n)$ is finitely determined then for k sufficiently large we shall have $\pi_k(f) \notin Z_k$.

It is easy to see that if $q \geq k$, then $\pi_{q,k}(Z_q) \subseteq Z_k$. For let $z \in Z_q$ and let $f \in m(n)$ represent z, i.e. suppose $z = \pi_q(f)$. If $\pi_{q,k}z \notin Z_k$, then $\tau_k(f) < k$ so by Corollary 2.16 $m(n)^k \subseteq \langle \partial f/\partial x_1, \ldots, \partial f/\partial x_n \rangle$.
Hence $\tau_k(f) = \tau(f)$, so since $q \geq k$, it follows that $\tau_q(f) = \tau_k(f) < k \leq q$. Hence $z \notin Z_q$, which is a contradiction. Therefore $\pi_{q,k}(z) \in Z_k$.

Proposition 2.22. Z_k is an algebraic subset of the Euclidean space $J_o^k(n,1)$.

Proof: Let $z \in J_o^k(n,1)$ and let $f \in m(n)$ represent z.

Now $z \in Z_k$ if and only if $\tau(z) \geq k$. But by definition,

$$\begin{aligned}\tau(z) = \tau_k(f) &= \dim_R \mathcal{E}(n)/(\langle \partial f/\partial x_1, \ldots, \partial f/\partial x_n \rangle + m(n)^k) \\ &= \dim_R J^{k-1}(n,1)/\pi_{k-1}\langle \partial f/\partial x_1, \ldots, \partial f/\partial x_n \rangle.\end{aligned}$$

So $z \in Z_k$ if and only if

$$(*) \quad \dim_{\mathbb{R}} \pi_{k-1}\langle \partial f/\partial x_1, \ldots, \partial f/\partial x_n \rangle \leqslant \dim_{\mathbb{R}} J^{k-1}(n,1)-k.$$

But $\pi_{k-1}\langle \partial f/\partial x_1, \ldots, \partial f/\partial x_n \rangle$ is generated over $\mathbb{R}$ by finitely many elements of the form $\pi_{k-1}(x^{\alpha}\partial f/\partial x_i)$, where α is an n-tuple of non-negative integers with $|\alpha| \leqslant k-1$; these elements all depend linearly on z, so (*) is evidently an algebraic condition on z. Hence Z_k is algebraic.

§ 3 UNIVERSAL UNFOLDINGS

In this section we shall be considering r-parameter families of germs in $\mathcal{E}(n)$ which contain a given germ of $m(n)$ at $0 \in \mathbb{R}^r$. To handle this situation conveniently we need some notational conventions.

We shall be considering germs in $\mathcal{E}(n+r)$ for some given n and r. We shall usually consider $\mathbb{R}^{n+r}$ to be factored as $\mathbb{R}^n \times \mathbb{R}^r$, so if we write, for example, "$(x,u) \in \mathbb{R}^{n+r}$", this should be understood as meaning "$x \in \mathbb{R}^n$ and $u \in \mathbb{R}^r$". Occasionally, too, we shall need to consider germs defined on $\mathbb{R}^{n+r+1}$, which we shall generally consider to be factored as $\mathbb{R}^n \times \mathbb{R}^r \times \mathbb{R}$ (so "$(x,u,t) \in \mathbb{R}^{n+r+1}$" will mean "$x \in \mathbb{R}^n$, $u \in \mathbb{R}^r$, and $t \in \mathbb{R}$"). We shall also apply this convention to germs of mappings. For example if $\Phi \in \mathcal{E}(n+s,n+r)$ and we write $\Phi = (\varphi,\psi)$, this will mean $\varphi \in \mathcal{E}(n+s,n)$, $\psi \in \mathcal{E}(n+s,r)$, and for all (x,u) near 0 in $\mathbb{R}^{n+s}$, $\Phi(x,u) = (\varphi(x,u), \psi(x,u)) \in \mathbb{R}^{n+r}$.

Unless otherwise stated we shall take coordinates $x_1,\ldots,x_n$ on $\mathbb{R}^n$ and coordinates $u_1,\ldots,u_r$ on $\mathbb{R}^r$.

We shall identify $\mathbb{R}^n$ with the subspace $\mathbb{R}^n \times \{0\}$ of $\mathbb{R}^{n+r}$. So if $f \in \mathcal{E}(n+r,p)$ we may speak of $f|\mathbb{R}^n \in \mathcal{E}(n,p)$. Also, for any p we shall consider $\mathcal{E}(r,p)$, to be imbedded in $\mathcal{E}(n+r,p)$ via π^*, where $\pi\colon \mathbb{R}^n \times \mathbb{R}^r \longrightarrow \mathbb{R}^r$ is the projection onto the second factor. So for example if $\psi \in \mathcal{E}(n+r,p)$, then to say ψ is in $\mathcal{E}(r,p)$ means that for $(x,u) \in \mathbb{R}^{n+r}$, $\psi(x,u)$ depends only on u, not on x; hence in this case we may write "$\psi(u)$"

as a convenient abbreviation for "$\phi(x,u)$".
In a similar way we may identify $\mathcal{E}(n)$ with a subring of $\mathcal{E}(n+r)$. In particular we may consider any germ in $\mathcal{E}(n)$ to be also an element of $\mathcal{E}(n+r)$.

Finally, we shall frequently wish to identify an arbitrary real number c with the element of $\mathcal{E}(n)$ whose value everywhere on $\mathbb{R}^n$ is c.

<u>Definition 3.1.</u> An <u>unfolding</u> is a pair (f,η) where $\eta \in m(n)$ and $f \in \mathcal{E}(n+r)$ and $f|\mathbb{R}^n = \eta$. The integer r is called the <u>unfolding dimension</u> of (f,η).

By abuse of language, we shall often speak of "the unfolding f" when we mean "the unfolding (f,η)", and we shall write "f is an (r-dimensional) unfolding of η" as an alternative for "(f,η) is an (r-dimensional) unfolding". Since f and r determine η uniquely, we may speak, without mentioning η, of a property which f has "as an r-dimensional unfolding" and we may even omit mention of r if it is clear from a previous statement what r is.

<u>Definition 3.2.</u> Let $\eta \in m(n)$. Let $f \in \mathcal{E}(n+r)$ and $g \in \mathcal{E}(n+s)$ be unfoldings of η.

A <u>right-left morphism</u> from f to g is a pair (Φ,λ), where $\Phi \in \mathcal{E}(n+r,n+s)$ and $\lambda \in \mathcal{E}(1+r)$, satisfying the following conditions:

(a) $\Phi|\mathbb{R}^n = id_{\mathbb{R}^n}$ (i.e. $\Phi(x,0) = (x,0)$ for all $x \in \mathbb{R}^n$, where 0 on the left is the origin of $\mathbb{R}^r$ and 0 on the right is the origin of $\mathbb{R}^s$).

(b) If $\Phi = (\varphi,\psi)$, where $\varphi \in \mathcal{E}(n+r,n)$ and $\psi \in \mathcal{E}(n+r,s)$, then $\psi \in \mathcal{E}(r,s)$.

(c) $\lambda|\mathbb{R} = id_{\mathbb{R}}$ (i.e. $\lambda(t,0) = t$ for all $t \in \mathbb{R}$; here 0 is the origin of $\mathbb{R}^r$).

(d) for all $(x,u) \in \mathbb{R}^{n+r}$

$$f(x,u) = \lambda(g(\Phi(x,u)),u).$$

Condition (b) says that Φ maps each "fibre" $\mathbb{R}^n\times\{u\} \subseteq \mathbb{R}^n\times\mathbb{R}^r$ ($u \in \mathbb{R}^r$) into some "fibre" $\mathbb{R}^n\times\{\psi(u)\} \subseteq \mathbb{R}^n\times\mathbb{R}^s$ and by condition (a) if u is near enough to 0 this map is in fact a germ of a local diffeomorphism. Note that we do <u>not</u> require that $\varphi|\{0\}\times\mathbb{R}^r = 0$, so these local diffeomorphisms do not necessarily preserve the origin of the "$\mathbb{R}^n$-fibres".

If (Φ,λ) is a right-left morphism from f to g, then by first choosing representatives we can for u near 0 in $\mathbb{R}^r$ define $\lambda_u \in \mathcal{E}(1)$ by $\lambda_u(t) = \lambda(t,u)$ for t near 0 in $\mathbb{R}$. Then condition (c) implies that λ_u is a germ of a local diffeomorphism of $\mathbb{R}$ if u is close to 0. And condition (d) can be rewritten as

$$f(x,u) = \lambda_u(g(\varphi(x,u),\psi(u)))$$

(or more concisely as

$$f_u = \lambda_u g_{\psi(u)} \varphi_u \ , \quad u \text{ near } 0 \text{ in } \mathbb{R}^n$$

where f_u, $g_{\psi(u)}$, φ_u are defined in the obvious way.)

If (Φ,λ) is a right-left morphism from f to g, and if λ has the additional property that for each u near 0 in $\mathbb{R}^r$, λ_u is just a <u>translation</u> of $\mathbb{R}$, then (Φ,λ) will be called a right morphism from f to g. More precisely:

<u>Definition 3.3.</u> Let $\eta \in m(n)$. Let $f \in \mathcal{E}(n+r)$ and $g \in \mathcal{E}(n+s)$ be unfoldings of η. A right-left morphism (Φ,λ) from f to g will be called a <u>right-morphism</u> from f to g if there is an element $\alpha \in m(r)$ such that for $(t,u) \in R\times R^r$, $\lambda(t,u) = t+\alpha(u)$.

Note that if (Φ,λ) is a right morphism from f to g and if α is as above, then condition 3.2(c) reduces to: $\alpha(0) = 0$, i.e. $\alpha \in m(r)$, and condition 3.2(d) reduces to:

(d') $f(x,u) = g(\Phi(x,u)) + \alpha(u)$ for $(x,u) \in R^{n+r}$.

If $\Phi \in \mathcal{E}(n+r,n+s)$ and if $\alpha \in m(r)$ we shall often write (Φ,α) to denote (Φ,λ), where $\lambda \in \mathcal{E}(1+r)$ is defined by $\lambda(t,u) = t+\alpha(u)$. Of course we shall only use this convention when (Φ,λ) is a right morphism.

It may not be clear why, if (Φ,λ) is a right morphism, we allow λ_u for each u to be a translation of R instead of requiring λ_u to be the identity mapping for each u, which would seem more consistent with the definition of right-equivalence of germs in § 2. However, there is no essential difference between the two theories. In allowing right morphisms to translate the values of the unfolding f in each R^n-fibre, we are in effect just ignoring the constant terms of the restrictions of f to the fibres. This is reasonable since we are interested in unfoldings as families of germs in $\mathcal{E}(n)$ and we have agreed that the constant term plays no role in determining the "structure" of such a germ. Moreover, our treatment turns out to give the simplest description of unfoldings, and finally, it has the advantage of being consistent with the treatment of Mather [10, Ch.3].

<u>Definition 3.4.</u> Let $\eta \in m(n)$ and let f and g be unfoldings of η. We say f is <u>right-left induced</u> from g if there is a right-left morphism from f to g and we say f is <u>right-induced from g</u> if there is a right morphism from f to g.

<u>Definition 3.5.</u> Let $\eta \in m(n)$ and let f be an unfolding of η. f is said to be a <u>right-left universal</u> unfolding of η if every unfolding of η can be right-left induced from f; f is said to be a <u>right-universal</u> unfolding of η if every unfolding of η can be right-induced from f.

Our main goal in this section will be an answer to the question "given $\eta \in m(n)$, what unfoldings can η have?" We shall answer this question (for most η) by giving a complete description of the universal unfoldings of η.

The definition of unfolding, the definitions of "right-morphism" and "right-universal", as well as all of the theory below dealing with right-universal unfoldings are due to Mather [10, Chapter III].

<u>Remark:</u> We point out that the unfoldings of a given germ $\eta \in m(n)$ can be made into a category in two ways: The right category of unfoldings of η has as objects the unfoldings of η and as morphisms the right morphisms between unfoldings; the right-left category of unfoldings of η again has as objects all unfoldings of η, but here the morphisms are all right-left

morphisms between unfoldings.

Let $\eta \in m(n)$, and let $f \in \mathcal{E}(n+r)$, $g \in \mathcal{E}(n+s)$, and $h \in \mathcal{E}(n+q)$ be unfoldings of η. Let (Φ,λ) be a right-left morphism from f to g and let (Φ',λ') be a right-left morphism from g to h. Suppose $\Phi = (\varphi,\psi)$ ($\varphi \in \mathcal{E}(n+r,n)$, $\psi \in \mathcal{E}(r,s)$). Then the composition of these morphisms is given by $(\Phi',\lambda')\circ(\Phi,\lambda) = (\Phi'',\lambda'')$, where $\Phi'' = \Phi'\circ\Phi \in \mathcal{E}(n+r,n+q)$ and where $\lambda'' \in \mathcal{E}(1+r)$ is defined by $\lambda''(t,u) = \lambda(\lambda'(t,\psi(u)),u)$, that is, $\lambda''_u = \lambda'_u\circ\lambda_{\psi(u)}$. Clearly (Φ'',λ'') is a right-left morphism from f to h. Moreover it is clear that if (Φ',λ') and (Φ,λ) are in fact right-morphisms, then so is their composition (Φ'',λ'').

Definition 3.6. Let $\eta \in m(n)$. Let $f \in \mathcal{E}(n+r)$ and $g \in \mathcal{E}(n+s)$ be unfoldings of η. Then f and g are said to be right (right-left) isomorphic if they are isomorphic objects of the right (right-left) category of unfoldings of η.

If f and g are right-isomorphic we shall write $f \cong_r g$; if they are right-left isomorphic we shall write $f \cong_{rl} g$.

Clearly a necessary condition for f and g to be isomorphic in either sense is that they have the same unfolding dimension, i.e., that $r = s$.

Suppose (Φ,λ) is a right-left morphism between two r-dimensional unfoldings of a given germ $\eta \in m(n)$, and suppose $\Phi = (\varphi,\psi)$, with $\varphi \in \mathcal{E}(n+r,n)$ and $\psi \in \mathcal{E}(r,r)$. Then it is easy to see that a necessary and sufficient condition for (Φ,λ) to be a right-left isomorphism is that $\phi \in L(r)$. Moreover it is clear that if (ϕ,λ) is a right morphism and has an inverse in the right-left category, then this inverse is also a right morphism. Hence a <u>right</u> morphism is a <u>right</u> isomorphism if and only if it is a <u>right-left</u> isomorphism.

<u>Definition 3.7.</u> Let $\eta \in m(n)$. The <u>r-dimensional constant unfolding</u> of η is the unfolding $f \in \mathcal{E}(n+r)$ defined by $f(x,u) = \eta(x)$ for $(x,u) \in R^{n+r}$.

<u>Definition 3.8.</u> Let $\eta \in m(n)$, and let $f \in \mathcal{E}(n+r)$ and $g \in \mathcal{E}(n+s)$ be unfoldings of η. Define an unfolding $f \oplus g \in \mathcal{E}(n+r+s)$ by $(f \oplus g)(x,u,v) = f(x,u)+g(x,v)-\eta(x)$ for $x \in R^n$, $u \in R^r$, $v \in R^s$. Clearly $f \oplus g$ is an r+s-dimensional unfolding of η, and can be considered as an s-dimensional unfolding of f, and also (after a permutation of coordinates) as an r-dimensional unfolding of g.

We now begin our investigation of universal unfoldings with a trivial but useful lemma:

<u>Lemma 3.9.</u> Let $\eta \in m(n)$. Let $f \in \mathcal{E}(n+r)$ be an unfolding of η. Note that any unfolding g of f can be considered also as an unfolding of η. Then

(a) f can be right-induced from any unfolding of f.

(b) Any constant unfolding of f can be right induced from f.

As a trivial corollary we have

(c) If g is any constant unfolding of f then

g is a right (right-left) universal unfolding of η if and only if f is.

Proof: Immediate from the definitions.

The main theorems of this section, which we shall state in a short while, will give a complete characterization of universal unfoldings in terms of certain algebraic conditions. We shall establish these theorems by showing that universality is equivalent to a certain transversality condition and we obtain the algebraic conditions simply by computation of the tangent spaces involved in this transversality condition. But first we must define the transversality condition we need.

Definition 3.10. Let $U \subseteq \mathbb{R}^{n+r}$ be an open neighbourhood of the origin. Let $g: U \longrightarrow \mathbb{R}$ be a smooth mapping. Let k be a non-negative integer. We shall define a smooth mapping $j_1^k g: U \longrightarrow J_0^k(n,1)$ as follows: For every $(x,u) \in U$ define a germ $g_{(x,u)} \in m(n)$ by setting $g_{(x,u)}(x') := g(x+x',u)-g(x,u)$ $(x' \in \mathbb{R}^n)$. Now define $j_1^k g: U \longrightarrow J_0^k(n,1)$ by setting $j_1^k g(x,u) := \pi_k(g_{(x,u)})$ for all $(x,u) \in U$.

From this definition it is clear that the germ at 0 of $j_1^k g$ depends only upon the germ at 0 of g. Hence we may use the same definition to associate to each germ $f \in \mathcal{E}(n+r)$ a germ $j_1^k f$ which is the germ at 0 of a smooth mapping from $\mathbb{R}^{n+r}$ into $J_0^k(n,1)$. Observe that if $\eta \in m(n)$ and if f in $\mathcal{E}(n+r)$ unfolds η, then $j_1^k f(0) = \pi_k(\eta)$.

We may restate this definition more concisely as follows: Let $i: \mathbb{R}^n \longrightarrow \mathbb{R}^{n+r}$ be the canonical linear imbedding (defined by $i(x) = (x,0)$ for $x \in \mathbb{R}^n$). Then if f is either a smooth real valued function defined on a neighbourhood of 0 in $\mathbb{R}^{n+r}$, or if f is a germ in $\mathcal{E}(n+r)$, we may define $j_1^k f$ by setting

$$j_1^k f = {}_k i^* \circ J_0^k f \qquad \text{(see definition 1.21).}$$

Remark: In this chapter we shall use this definition only for germs.

<u>Definition 3.11.</u> Let $\eta \in m(n)$ and let $f \in \mathcal{E}(n+r)$ be an unfolding of η. Let k be a non-negative integer and let $z = \pi_k(\eta) \in J_0^k(n,1)$.

We say f is <u>right k-transversal</u> if $j_1^k f$ is transversal at 0 to $zL^k(n)$.

We say f is <u>right-left k-transversal</u> if $j_1^k f$ is transversal at 0 to $L^k(1)zL^k(n)$.

Our next lemma equates k-transversality to an algebraic condition.

<u>Definition 3.12.</u> Let $\eta \in m(n)$, and let $f \in \mathcal{E}(n+r)$ be an unfolding of η.

For $1 \leq i \leq r$, set $\alpha_i(f) = \partial f/\partial u_i | \mathbb{R}^n \in \mathcal{E}(n)$. Denote by 1 the multiplicative identity of $\mathcal{E}(n)$; i.e. 1 is the germ at 0 of the function whose value everywhere on $\mathbb{R}^n$ is $1 \in \mathbb{R}$.

Then let $V_f = \langle 1, \alpha_1, \dots, \alpha_r \rangle_{\mathbb{R}}$, the linear subspace of $\mathcal{E}(n)$ generated over $\mathbb{R}$ by 1 and by the α_i's.

Let $W_f = \langle \alpha_1, \dots, \alpha_r \rangle_{\mathbb{R}}$. W_f is a linear subspace of V_f.

<u>Lemma 3.13.</u> Let $\eta \in m(n)$ and let $f \in \mathcal{E}(n+r)$ be an unfolding of η. Let k be a non-negative integer. Then

(a) f is right k-transversal if and only if

$$\mathcal{E}(n) = \langle \partial\eta/\partial x_1, \dots, \partial\eta/\partial x_n \rangle + V_f + m(n)^{k+1}$$

(b) f is right-left k-transversal if and only if

$$\mathcal{E}(n) = \langle \partial\eta/\partial x_1, \dots, \partial\eta/\partial x_n \rangle + \eta^*\mathcal{E}(1) + W_f + m(n)^{k+1}$$

(Note, by the way, that $\eta^*\mathcal{E}(1)+W_f = \eta^*\mathcal{E}(1)+V_f$, since $1 \in \eta^*\mathcal{E}(1)$).

Proof: Since $\mathbb{R}^{n+r}$ and $J_o^k(n,1)$ are Euclidean spaces, we may identify the tangent space of $\mathbb{R}^{n+r}$ at 0 with $\mathbb{R}^{n+r}$ and we may identify the tangent space of $J_o^k(n,1)$ at $\pi_k(\eta)$ with $J_o^k(n,1)$ itself. Associated to $j_1^k f$ there is a linear map $Tj_1^k f\colon \mathbb{R}^{n+r} \longrightarrow J_o^k(n,1)$ of these tangent spaces (recall $j_1^k f(0) = \pi_k(\eta)$).

Now the tangent space to $\mathbb{R}^{n+r}$ at 0 is generated over $\mathbb{R}$ by the tangent vectors $\frac{\partial}{\partial x_1}, \dots, \frac{\partial}{\partial x_n}, \frac{\partial}{\partial u_1}, \dots, \frac{\partial}{\partial u_r}$.

We recall the definition of $j_1^k f$. Pick a representative f' of f defined on an neighbourhood U of 0 in $\mathbb{R}^{n+r}$ and for (x,u) in U define $f'_{(x,u)} \in m(n)$ by $f'_{(x,u)}(x') = f'(x+x',u) - f'(x,u)$ for $x' \in \mathbb{R}^n$. Then $j_1^k f(x,u) = \pi_k(f'_{(x,u)})$.

So if $1 \leq i \leq n$, $Tj_1^k f(\frac{\partial}{\partial x_i}) = \frac{\partial}{\partial x_i} j_1^k f \mid_{(x,u)=0}$

$$= \frac{\partial}{\partial x_i}(\pi_k(f'_{(x,u)}))\mid_{(x,u)=0}$$

$$= \pi_k\left(\frac{\partial}{\partial x_i}(f'_{(x,u)})\mid_{(x,u)=0}\right).$$

Now $\left(\frac{\partial}{\partial x_i} f'_{(x,u)}\mid_{(x,u)=0}\right)(x') = \frac{\partial}{\partial x_i}(f'(x+x',u)-f'(x,u))\mid_{(x,u)=0}$

$= \frac{\partial f'}{\partial x_i}(x',0) - \frac{\partial f'}{\partial x_i}(0,0)$. This has the same germ at 0 as

$\frac{\partial \eta}{\partial x_i} - \frac{\partial \eta}{\partial x_i}(0)$. So $Tj_1^k f\left(\frac{\partial}{\partial x_i}\right) = \pi_k\left(\frac{\partial \eta}{\partial x_i} - \frac{\partial \eta}{\partial x_i}(0)\right)$.

To simplify the notation we set $\gamma_i = \frac{\partial \eta}{\partial x_i} - \frac{\partial \eta}{\partial x_i}(0)$ for $i = 1,\ldots,n$.

By a similar computation, if $1 \leq j \leq r$, $Tj_1^k f\left(\frac{\partial}{\partial u_j}\right)$

$= \pi_k\left(\frac{\partial}{\partial u_j}(f'_{(x,u)})\mid_{(x,u)=0}\right)$ and

$\left(\frac{\partial}{\partial u_j}(f'_{(x,u)})\mid_{(x,u)=0}\right)(x') = \frac{\partial}{\partial u_j}(f'(x+x',u)-f'(x,u))\mid_{(x,u)=0}$

$= \frac{\partial f'}{\partial u_j}(x',0) - \frac{\partial f'}{\partial u_j}(0,0)$. This has the same germ at 0 as

$\alpha_j(f) - \frac{\partial f}{\partial u_j}(0,0)$. To simplify the notation below, if $1 \leq j \leq r$

we set $\beta_j = \alpha_j(f) - \frac{\partial f}{\partial u_j}(0,0)$.

Let $z = \pi_k(\eta)$. By lemma 2.8 $T_z\ zL^k(n) = \pi_k(m(n)\langle \partial\eta/\partial x_1,\ldots,\partial\eta/\partial x_n\rangle$ and f is right k-transversal if and only if this space and the image of $Tj_1^k f$ together span $J_o^k(n,1)$, i.e. if and only if $\pi_k(m(n)\langle \partial\eta/\partial x_1,\ldots,\partial\eta/\partial x_n\rangle) + \langle \pi_k(\gamma_1),\ldots,\pi_k(\gamma_n),\pi_k(\beta_1),\ldots,\pi_k(\beta_r)\rangle$ $= J_o^k(n,1) = \pi_k(m(n))$.

But this holds if and only if

$$m(n)\langle \partial\eta/\partial x_1,\ldots,\partial\eta/\partial x_n\rangle + \langle \gamma_1,\ldots,\gamma_n,\beta_1,\ldots,\beta_r\rangle_R + m(n)^{k+1} = m(n),$$

or equivalently if and only if

$$(*)\quad \mathcal{E}(n) = \langle \partial\eta/\partial x_1,\ldots,\partial\eta/\partial x_n\rangle_{m(n)} + \langle 1,\gamma_1,\ldots,\gamma_n,\beta_1,\ldots,\beta_r\rangle_R + m(n)^{k+1}.$$

But clearly $\langle 1,\gamma_1,\ldots,\gamma_n,\beta_1,\ldots,\beta_r\rangle_R =$

$$= \langle 1,\partial\eta/\partial x_1,\ldots,\partial\eta/\partial x_n,\alpha_1(f),\ldots,\alpha_r(f)\rangle_R$$

$$= \langle \partial\eta/\partial x_1,\ldots,\partial\eta/\partial x_n\rangle_R + V_f .$$

So $(*)$ is equivalent to: $\mathcal{E}(n) = \langle \partial\eta/\partial x_1,\ldots,\partial\eta/\partial x_n\rangle + V_f + m(n)^{k+1}$ which proves (a).

Again by 2.8 we have $T_z L^k(1) z L^k(n) = \pi_k\big(m(n)\langle \partial\eta/\partial x_1,\ldots,\partial\eta/\partial x_n\rangle + \eta^* m(1)\big)$.

So by a similar argument to the one above, f is right-left k-transversal if and only if

$$m(n) = m(n)\langle \partial\eta/\partial x_1,\ldots,\partial\eta/\partial x_n\rangle + \eta^* m(1) + \langle \gamma_1,\ldots,\gamma_n,\beta_1,\ldots,\beta_r\rangle_R + m(n)^{k+1}$$

or equivalently, if and only if

$$(**)\quad \mathcal{E}(n) = m(n)\langle \partial\eta/\partial x_1,\ldots,\partial\eta/\partial x_n\rangle + \eta^* m(1) + \langle 1,\gamma_1,\ldots,\gamma_n,\beta_1,\ldots,\beta_r\rangle_R + m(n)^{k+1} .$$

But again $\langle 1,\gamma_1,\ldots,\gamma_n,\beta_1,\ldots,\beta_r\rangle_R =$

$$\langle 1,\partial\eta/\partial x_1,\ldots,\partial\eta/\partial x_n,\alpha_1(f),\ldots,\alpha_r(f)\rangle_R$$

$$= \langle \partial\eta/\partial x_1,\ldots,\partial\eta/\partial x_n\rangle_R + \langle 1\rangle_R + W_f .$$

And since $\eta^* m(1) + \langle 1\rangle_R = \eta^*\mathcal{E}(1)$ it follows that $(**)$ is equivalent to

$$\mathcal{E}(n) = \langle \partial\eta/\partial x_1, \ldots, \partial\eta/\partial x_n \rangle + \eta^* \mathcal{E}(1) + W_f + m(n)^{k+1} ,$$

which proves (b).

Corollary 3.14. Let $\eta \in m(n)$. Then for every non-negative integer k one can find an unfolding f of η which is right (and hence also right-left) k-transversal.

Proof: Let k be a non-negative integer.
$\mathcal{E}(n)/(\langle \partial\eta/\partial x_1, \ldots, \partial\eta/\partial x_n \rangle + m(n)^{k+1})$ is a finite-dimensional $\mathbb{R}$-vector space; choose elements $b_1, \ldots, b_r$ of $\mathcal{E}(n)$ whose cosets modulo the "denominator" generate this vector space over $\mathbb{R}$. Define $f \in \mathcal{E}(n+r)$ by

$$f(x,u) = \eta(x) + u_1 b_1(x) + \ldots + u_r b_r(x) .$$

Clearly f unfolds η. Moreover $\frac{\partial f}{\partial u_i} \Big| \mathbb{R}^n = b_i$ for $1 \leq i \leq r$, so the b_i generate $W_f \subseteq V_f$. Hence by the choice of the b_i it is immediate from Lemma 3.13 that f is right k-transversal.

It is obvious from Lemma 3.13 that any right k-transversal unfolding of η is also right-left k-transversal.

Corollary 3.15. Let $\eta \in m(n)$ and let $f \in \mathcal{E}(n+r)$ be a right (right-left) universal unfolding of η. Then f is right (right-left) k-transversal for any non-negative integer k.

Proof: We shall prove both cases simultaneously. Let a non-negative integer k be given. By Corollary 3.14 we can find a right k-transversal unfolding g of η, with $g \in \mathcal{E}(n+s)$ for some s.

Now suppose f is right or right-left universal. Then f is in any event right-left universal, so we can find a right-left morphism (Φ,λ) from g to f. By the definition of "right-left morphism", if $\Phi = (\varphi,\psi)$ then $\psi \in \mathcal{E}(s,r)$ and we have

$$g(x,v) = \lambda(f(\varphi(x,v),\psi(v)),v) \text{ for } (x,v) \in R^{n+s}.$$

Hence if $1 \leq \ell \leq s$ we have, for $x \in R^n$:

$$(*)\ \alpha_\ell(g)(x) = \frac{\partial g}{\partial v_\ell}(x,0) = \frac{\partial\lambda}{\partial t}(f(x,0),0)\Big(\sum_{i=1}^{n} \frac{\partial f}{\partial x_i}(x,0)\,\frac{\partial\varphi_i}{\partial v_\ell}(x,0)$$

$$+ \sum_{j=1}^{r} \frac{\partial f}{\partial u_j}(x,0)\,\frac{\partial\psi_j}{\partial v_\ell}(0)\Big) + \frac{\partial\lambda}{\partial v_\ell}(f(x,0),0)$$

$$= \sum_{i=1}^{n} \frac{\partial\eta}{\partial x_i}(x)\,\frac{\partial\varphi_i}{\partial v_\ell}(x,0) + \sum_{j=1}^{r} \alpha_j(f)(x)\,\frac{\partial\psi_j}{\partial v_\ell}(0)$$

$$+ \frac{\partial\lambda}{\partial v_\ell}(\eta(x),0)$$

(recall that $\varphi(x,0) = x$, $\psi(0) = 0$, and $\lambda(t,0) = 0$ (so $\frac{\partial\lambda}{\partial t}(t,0) = 1$), for $x \in R^n$, $t \in R$).

So clearly $W_g \subseteq \langle \partial\eta/\partial x_1,\dots,\partial\eta/\partial x_n\rangle + W_f + \eta^*\mathcal{E}(1)$ and since g is right left k-transversal, so is f, by Lemma 3.13.

Moreover if f is right universal, we can pick (Φ,λ) to be a right morphism. That is, there is an $\alpha \in m(s)$ so that

$$\lambda(t,v) = t+\alpha(v) \qquad ((t,v) \in R^{1+s})$$

and hence $\partial\lambda/\partial v_\ell(t,0) = \partial\alpha/\partial v_\ell(0)$ and is constant as a function of t. So $\partial\lambda/\partial v_\ell(\eta,0)$ is a constant germ, i.e. a real multiple of $1 \in \mathcal{E}(n)$, and hence by (*), $V_g \subseteq \langle \partial\eta/\partial x_1,\dots,\partial\eta/\partial x_n\rangle + V_f$. Hence by Lemma 3.13, since g is right k-transversal so is f. This completes the proof.

We now state the main lemma of this section, which will enable us to give a complete characterization of universal unfoldings:

Lemma 3.16. Let k be a non-negative integer, and suppose $\eta \in m(n)$ is right (right-left) k-determined.

If $f \in \mathcal{E}(n+r)$ and $g \in \mathcal{E}(n+r)$ are right (right-left) k-transversal unfoldings of η of the same unfolding dimension, then f and g are right (right-left) isomorphic.

We postpone the proof till the end of the chapter, and consider first the consequences of this lemma.

Theorem 3.17. Let k be a non-negative integer, and suppose $\eta \in m(n)$ is right (right-left) k-determined.

Then an unfolding f of η is right (right-left) universal if and only if f is right (right-left) k-transversal.

Proof: "Only if" is trivial by Corollary 3.15.

So suppose now $f \in \mathcal{E}(n+r)$ is a right (right-left) k-transversal unfolding of η. Let $g \in \mathcal{E}(n+s)$ be any unfolding of η. We must show that g can be right (right-left) induced from f. Now consider the unfolding $g \oplus f \in \mathcal{E}(n+s+r)$. Obviously (g, η) can be right induced from $(g \oplus f, \eta)$, since $g \oplus f$ unfolds g. And $g \oplus f$ is obviously a right (right-left) k-transversal unfolding of η, since f is. For the same reason, if $h \in \mathcal{E}(n+r+s)$ is the s-dimensional constant unfolding of f, then h is a right (right-left) k-transversal unfolding of η. So by Lemma 3.16, there is a right (right-left) isomorphism from $(g \oplus f, \eta)$ to (h, η). Moreover, (f, η) right induces (h, η). So we have morphisms from g to $g \oplus f$ to h to f; composing them gives a right (right-left) morphism from g to f.

Lemma 3.18. Let $\eta \in m(n)$ and let $f \in \mathcal{E}(n+r)$ unfold η. If for every k, the unfolding f is right (right-left) k-transversal, then f is right (right-left) universal and η is finitely determined.

Proof: By Theorem 3.17 all we must show is that under the hypotheses, η is finitely determined. But if f is right or right-left k-transversal then in any event f is certainly right-left k-transversal, so by Lemma 3.13 we have for every non-negative integer k

$$\mathcal{E}(n) = \langle \partial\eta/\partial x_1, \ldots, \partial\eta/\partial x_n \rangle + \eta^* \mathcal{E}(1) + W_f + m(n)^{k+1}$$

and hence $\sigma_{k+1}(\eta) \leqslant \dim_R W_f \leqslant r$. So $\sigma_k(\eta)$ is bounded and hence by Corollary 2.17 η is finitely determined.

Theorem 3.19. (Existence of universal unfoldings) Let $\eta \in m(n)$. Then η has a right (right-left) universal unfolding if and only if η is finitely determined.

Proof: "If" follows immediately from Theorem 3.17 and Corollary 3.14. "Only if" is a consequence of Corollary 3.15 and Lemma 3.18.

Theorem 3.20. (Uniqueness of universal unfoldings)
If $\eta \in m(n)$, then any two right (right-left) universal unfoldings of η of the same dimension are right (right-left) isomorphic.

Proof: Let $f \in \mathcal{E}(n+r)$ and $g \in \mathcal{E}(n+r)$ be right (right-left) universal unfoldings of η. Then η must be finitely determined, and f and g are right (right-left) k-transversal for every non-negative integer k by Corollary 3.15, so the result follows trivially by Lemma 3.16.

Corollary 3.21. Let $\eta \in m(n)$. Let $f \in \mathcal{E}(n+r)$ and $g \in \mathcal{E}(n+s)$ be right (right-left) universal unfoldings of η, and suppose $s \geq r$. Then g is right (right-left) isomorphic (as unfoldings of η) to a constant unfolding of f.

Proof: This is trivial by theorem 3.20, since the constant unfolding of f of any dimension is also a right (right-left) universal unfolding of η.

Theorem 3.22. Let $\eta \in m(n)$ and let $f \in \mathcal{E}(n+r)$ unfold η. Then f is right-universal if and only if

(a) $\quad \mathcal{E}(n) = \langle \partial\eta/\partial x_1, \ldots, \partial\eta/\partial x_n \rangle + V_f$

and f is right-left universal if and only if

(b) $\quad \mathcal{E}(n) = \langle \partial\eta/\partial x_1, \ldots, \partial\eta/\partial x_n \rangle + \eta^*\mathcal{E}(1) + W_f$.

Proof: "If". If (a) (resp. (b)) holds then by Lemma 3.13, f is right (right-left) k-transversal for every non-negative integer k, so by Lemma 3.18, f is right (right-left) universal.

"Only if". If f is right (right-left) universal then η is finitely determined, so for some k ($0 \leq k < \infty$) we have $m(n)^{k+1} \subseteq \langle \partial\eta/\partial x_1, \ldots, \partial\eta/\partial x_n \rangle$. Moreover for this k f is right (right-left) k-transversal by Corollary 3.15. Then (a) (resp. (b)) follows from Lemma 3.13.

<u>Theorem 3.23.</u> Let $\eta \in m(n)^2$ be finitely determined. Then the minimal dimension of a right (right-left) universal unfolding of η is the right (right-left) codimension of η.

In fact if $b_1, \ldots, b_r \in m(n)$ are chosen so that their classes in $m(n)/\langle \partial\eta/\partial x_1, \ldots, \partial\eta/\partial x_n \rangle$ form a basis of this $\mathbb{R}$-vector space, and if we define $f \in \mathcal{E}(n+r)$ by $f(x,u) = \eta(x)+u_1b_1(x)+\ldots+u_rb_r(x)$, then f is a right universal unfolding of η of minimal dimension.

And if $a_1, \ldots, a_s \in m(n)$ are chosen so that their classes in $m(n)/(\langle \partial\eta/\partial x_1, \ldots, \partial\eta/\partial x_n \rangle + \eta^* m(1))$ are a basis of this $\mathbb{R}$-vector space, and if we define $g \in \mathcal{E}(n+s)$ by $g(x,v) = \eta(x) + v_1a_1(x) + \ldots + v_sa_s(x)$, then g is a right-left universal unfolding of η of minimal dimension.

<u>Proof:</u> Let $h \in \mathcal{E}(n+q)$ be an unfolding of η. If h is right-universal then by 3.22(a) we see immediately that $\tau(\eta) \leq \dim_{\mathbb{R}} V_h \leq q+1$. Since r-codim $\eta = \tau(\eta)-1$ we have $q \geq$ r-codim(η).

If h is right-left universal then by 3.22(b) we see that rl-codim$(\eta) = \sigma(\eta) \leq \dim_{\mathbb{R}} W_h \leq q$.

So the dimension of a right (right-left) universal unfolding of η is at least the right (right-left) codimension of η.

Now suppose we choose $b_1,\ldots,b_r \in m(n)$ so that their classes span $m(n)/\langle\partial\eta/\partial x_1,\ldots,\partial\eta/\partial x_n\rangle$, and define $f \in \mathcal{E}(n+r)$ as above. Clearly r is just the right-codimension of η. And clearly $\alpha_i(f) = b_i$ for $1 \leqslant i \leqslant r$ so the b_i generate W_f. Hence $m(n) = \langle\partial\eta/\partial x_1,\ldots,\partial\eta/\partial x_n\rangle + W_f$. So since $V_f = W_f + \langle 1\rangle_R$ we have $\mathcal{E}(n) = \langle\partial\eta/\partial x_1,\ldots,\partial\eta/\partial x_n\rangle + V_f$ and hence by Theorem 3.22 f is right-universal.

Similarly if $a_1,\ldots,a_s$ are chosen and g is defined as in the statement of the theorem then s must be the right-left codimension of η, and clearly $\alpha_j(g) = a_j$ for $j = 1,\ldots,s$. Hence the a_j generate W_g and because of the way the a_j were chosen $\mathcal{E}(n) = \langle\partial\eta/\partial x_1,\ldots,\partial\eta/\partial x_n\rangle + \eta^*\mathcal{E}(1) + W_g$. So by Theorem 3.22 g is right-left universal. This completes the proof.

<u>Remark:</u> If $\eta \in m(n)$ and $\eta \notin m(n)^2$ then η is a right universal unfolding of itself. This is clear from Theorem 3.22, since some $\partial\eta/\partial x_i$ is a unit in $\mathcal{E}(n)$ and hence $\mathcal{E}(n) = \langle\partial\eta/\partial x_1,\ldots,\partial\eta/\partial x_n\rangle$.

<u>Remark:</u> If $\eta \in m(n)$ is finitely determined, then η has right (right-left) universal unfoldings of all dimensions greater than the minimal dimension, for any constant unfolding of a right (right-left) universal unfolding of minimal dimension is again a right (right-left) universal unfolding of η.

All that remains to be done now in this chapter is to prove Lemma 3.16.

<u>Proof of Lemma 3.16.</u> As far as is possible we shall for conciseness treat the right and right-left cases simultaneously.

Let a non-negative integer k be given, let $\eta \in m(n)$ be right (right-left) k-determined, and let f and $g \in \mathcal{E}(n+r)$ be r-dimensional right (right-left) k-transversal unfoldings of η. We must construct a right (right-left) isomorphism from f to g.

We show first that it is enough to do this in the special case in which for every real number $s \in [0,1]$ the unfolding $F_s = sg + (1-s)f$ is also right (right-left) k-transversal.

(If this case holds for f and g we shall say f and g are right (right-left) elementarily k-homotopic; we shall say f and g are right (right-left) k-homotopic if we can find a finite sequence of unfoldings $f=f_0, f_1, \dots, f_{p-1}, f_p=g$ such that f_i and f_{i+1} are right (right-left) elementarily k-homotopic for $i = 0, \dots, p-1$).

Let $A \subseteq J^k(n,1)$ be defined as follows:
Set $A = \pi_k(\langle \partial\eta/\partial x_1, \dots, \partial\eta/\partial x_n \rangle + \langle 1 \rangle_R)$ if we are considering the "right" case and set $A = \pi_k(\langle \partial\eta/\partial x_1, \dots, \partial\eta/\partial x_n \rangle + \eta^* \mathcal{E}(1))$ if we are considering the "right-left" case. Now if $h \in \mathcal{E}(n+r)$ is any unfolding of η then by Lemma 3.13 h is right (right-left) k-transversal if and only if $\pi_k(W_h)$ is transversal to the linear subspace A of $J^k(n,1)$.

We now make two remarks. The first is that if $b_1, \dots, b_r$ are any r elements of $J^k(n,1)$, then η has an unfolding $h' \in \mathcal{E}(n+r)$ such that $\pi_k(\alpha_i(h')) = b_i$ for $i = 1, \dots, r$. Simply choose elements $\mu_1,\ldots,\mu_r$ in $\mathcal{E}(n)$ with $\pi_k(\mu_i) = b_i$ for $i = 1,\ldots,r$, and define $h'(x,u) = \eta(x)+u_1\mu_1(x)+\ldots+u_r\mu_r(x)$.

The second remark is the following. Let $h \in \mathcal{E}(n+r)$ be any right (right-left) k-transversal unfolding of η. Let i be an integer, $1 \leq i \leq r$, and let $b \in J^k(n,1)$ have the form

$$b = a + \sum_{j=1}^{r} c_j\pi_k\alpha_j(h), \quad \text{where } a \in A,\ c_j \in \mathbb{R} \text{ and } c_i > 0.$$

Choose any unfolding $h' \in \mathcal{E}(n+r)$ of η such that $\pi_k\alpha_i(h') = b$ and such that $\pi_k\alpha_j(h') = \pi_k\alpha_j(h)$ if $j \neq i$. (By the first remark such an h' always exists). Then h' is also right (right-left) k-transversal and moreover is right (right-left) elementarily k-homotopic to h, for since $c_i > 0$ one easily computes that $\pi_k W_{sh'+(1-s)h}$ is transversal to A for every $s \in [0,1] \subseteq \mathbb{R}$.

Now recall that we have a standard linear isomorphism of $J^k(n,1)$ to a Euclidean space, hence also a standard inner product on $J^k(n,1)$, so that we may speak of the orthogonal space to A, $A^{\perp}$. Let $\dim_{\mathbb{R}} A^{\perp} = q$, and choose a basis $z_1,\ldots,z_q$ of $A^{\perp}$. Let $h \in \mathcal{E}(n+r)$ be a right (right-left) k-transversal unfolding of η. Now by elementary linear algebra it is clear that by successive applications of the second remark above h is right (right-left) k-homotopic to an unfolding $h' \in \mathcal{E}(n+r)$ of η, having the property that (if $q > 0$) $\pi_k(\alpha_1(h')) = \pm z_1$; if $1 < j \leq q$ then $\pi_k(\alpha_j(h')) = z_j$; and if $q < j \leq r$ then $\pi_k(\alpha_j(h')) = 0$ (an unfolding h' with this property will be called right (right-left) k-orthogonal).

In particular, our given unfoldings f and g are right (right-left) k-homotopic to right (right-left) k-orthogonal unfoldings f' and g' respectively. Now if $q > 0$ and if $\pi_k(\alpha_1(f')) = -\pi_k(\alpha_1(g'))$ define $f'' \in \mathcal{E}(n+r)$ by

$$f''(x_1,\dots,x_n,u_1,\dots,u_r) = f'(x_1,\dots,x_n,-u_1,u_2,\dots,u_r);$$

otherwise set $f'' = f'$. Clearly f" and f' are right isomorphic unfoldings of η, and for all j ($1 \leq j \leq r$) we have $\pi_k(\alpha_j(f'')) = \pi_k(\alpha_j(g'))$. So by our second remark above f" is right (right-left) elementarily k-homotopic to g'.

So we see that if $f \in \mathcal{E}(n+r)$ and $g \in \mathcal{E}(n+r)$ are any right (right-left) k-transversal unfoldings of η, of the same dimension, then we can find r-dimensional right (right-left) k-transversal unfoldings of η:

$f=f_0,f_1,\dots,f_{p-1},f_p=g$ such that f_i and f_{i+1} are either right (right-left) elementary k-homotopic or right isomorphic, for $i = 1,\dots,p$. Hence it is enough to prove the lemma in the case when f and g are right (right-left) elementarily k-homotopic.

So suppose this is the case.For $s \in \mathbb{R}$, define an unfolding $F_s \in \mathcal{E}(n+r)$ of η by setting $F_s = (1-s)f + sg$. We shall show that for every $s \in [0,1]$, if s' is sufficiently close to s then $F_{s'}$ is right (right-left) isomorphic to F_s. Clearly (either by the compactness or by the connectedness of the interval [0,1]) this implies that $f = F_0$ is right (right-left) isomorphic to $F_1 = g$.

So let $s \in [0,1]$ be given. Define $H \in \mathcal{E}(n+r+1)$ by setting $H(x,u,t) = F_{s+t}(x,u) = (1-s-t)f(x,u)+(s+t)g(x,u)$ for $(x,u) \in \mathbb{R}^{n+r}$, $t \in \mathbb{R}$. The idea of the proof now is to construct isomorphisms between F_s and $F_{s'}$, for s' close to s, by applying Lemma 1.29, one of our propositions on solutions of differential equations, to H. But first we must perform some "algebraic" manipulations in order to derive a suitable form of the hypotheses of Lemma 1.29.

Since by assumption F_s is right (right-left) k-transversal, we have by Lemma 3.13

(a)

$\mathcal{E}(n) = \langle \partial\eta/\partial x_1, \ldots, \partial\eta/\partial x_n \rangle + \langle 1 \rangle_{\mathbb{R}} + W_{F_s} + m(n)^{k+1}$ in the "right" case and

$\mathcal{E}(n) = \langle \partial\eta/\partial x_1, \ldots, \partial\eta/\partial x_n \rangle + \eta^*\mathcal{E}(1) + W_{F_s} + m(n)^{k+1}$ in the "right-left" case.

We shall show that in fact the following stronger equations hold

(b)

$\mathcal{E}(n) = \langle \partial\eta/\partial x_1, \ldots, \partial\eta/\partial x_n \rangle + \langle 1 \rangle_{\mathbb{R}} + W_{F_s}$ in the "right" case;

$\mathcal{E}(n) = \langle \partial\eta/\partial x_1, \ldots, \partial\eta/\partial x_n \rangle + \langle 1, \eta, \eta^2, \ldots, \eta^{k+1} \rangle_{\mathbb{R}} + W_{F_s}$

in the "right-left" case.

In the "right" case this is easy to see, for since η is right k-determined, Corollary 2.10 tells us that $m(n)^{k+1} \subset m(n)\langle \partial\eta/\partial x_1, \ldots, \partial\eta/\partial x_n \rangle$. The "right-left" case requires more work, because we wish not only to eliminate the term $m(n)^{k+1}$ from the second equation (a), but also wish to replace the term $\eta^*\mathcal{E}(1)$ by something finitely generated over $\mathbb{R}$. (This is necessary for a later step in the proof). First observe that since η is right-left k-determined, we have by Corollary 2.10

(c) $\mathfrak{m}(n)^{k+1} \subseteq \mathfrak{m}(n)\langle \partial\eta/\partial x_1, \ldots, \partial\eta/\partial x_n\rangle + \eta^*\mathfrak{m}(1)$.

So we may first omit the term $\mathfrak{m}(n)^{k+1}$ from the second equation (a). Now $\eta^*\mathcal{E}(1) = \langle 1, \eta, \eta^2, \ldots, \eta^{k+1}\rangle_{\mathbb{R}} + \eta^*(\mathfrak{m}(1)^{k+2})$, and since $\eta \in \mathfrak{m}(n)$ it follows that $\eta^*(\mathfrak{m}(1)^{k+2}) \subseteq \mathfrak{m}(n)^{k+2}$. So from the second equation (a) we can derive

(d) $\mathcal{E}(n) = \langle \partial\eta/\partial x_1, \ldots, \partial\eta/\partial x_n\rangle + \langle 1, \eta, \ldots, \eta^{k+1}\rangle_{\mathbb{R}} + W_{F_s} + \mathfrak{m}(n)^{k+2}$.

But $\mathfrak{m}(n)^{k+2} \subseteq \langle \partial\eta/\partial x_1, \ldots, \partial\eta/\partial x_n\rangle$. (If $\eta \in \mathfrak{m}(n)^2$ this follows from Theorem 1.19 applied to (c), and if $\eta \notin \mathfrak{m}(n)^2$ this is trivial since then some $\partial\eta/\partial x_i$ is a unit, so $\langle \partial\eta/\partial x_1, \ldots, \partial\eta/\partial x_n\rangle = \mathcal{E}(n)$.) Hence we may omit the term $\mathfrak{m}(n)^{k+2}$ in (d) above; this then yields the second equation (b).

Now let h be any germ in $\mathcal{E}(n+r+1)$. Then $h|\mathbb{R}^n \in \mathcal{E}(n)$, so by (b) we may find elements $\mu_1, \ldots, \mu_n$ of $\mathcal{E}(n)$, elements $c_o, \ldots, c_{k+1}$ of $\mathbb{R}$ (with $c_1 = c_2 = \ldots = c_{k+1} = 0$ in the "right" case) and elements $b_1, \ldots, b_r$ of $\mathbb{R}$ such that

(e) $h|\mathbb{R}^n = \sum_{i=1}^{n} \frac{\partial\eta}{\partial x_i}\mu_i + \sum_{j=o}^{k+1} c_j\eta^j + \sum_{\ell=1}^{r} b_\ell\alpha_\ell(F_s)$.

If we consider the μ_i to be elements of $\mathcal{E}(n+r+1)$, we may define an element $h' \in \mathcal{E}(n+r+1)$ by setting

$$h' = h - \sum_{i=1}^{n} \frac{\partial H}{\partial x_i}\mu_i - \sum_{j=o}^{k+1} c_j H^j - \sum_{\ell=1}^{r} b_\ell \frac{\partial H}{\partial u_\ell} .$$

Now observe that $H|\mathbb{R}^{n+r} = F_s$ and $H|\mathbb{R}^n = \eta$; hence it clearly follows from (e) that $h'|\mathbb{R}^n = 0$. But this means $h' \in m(r+1)\mathcal{E}(n+r+1)$, by Lemma 1.4. Define $B \subseteq \mathcal{E}(n+r+1)$ by setting $B = \langle 1\rangle_{\mathbb{R}}$ in the "right" case and setting $B = \langle 1, H, H^2, \ldots, H^{k+1}\rangle_{\mathbb{R}}$ in the "right-left" case. Since the argument above is valid for any $h \in \mathcal{E}(n+r+1)$ we have shown:

$$\mathcal{E}(n+r+1) = \langle \partial H/\partial x_1, \ldots, \partial H/\partial x_n\rangle_{\mathcal{E}(n)} + B + \langle \partial H/\partial u_1, \ldots, \partial H/\partial u_r\rangle_{\mathbb{R}} + m(r+1)\mathcal{E}(n+r+1) .$$

Now it does not hurt the validity of the equation if we make the terms on the right bigger, so we may write:

$$\mathcal{E}(n+r+1) = \langle \partial H/\partial x_1, \ldots, \partial H/\partial x_n\rangle_{\mathcal{E}(n+r+1)} + \langle \partial H/\partial u_1, \ldots, \partial H/\partial u_r\rangle_{\mathcal{E}(r+1)} + B\mathcal{E}(r+1) + m(r+1)\mathcal{E}(n+r+1).$$

Now since B is a finite dimensional vector space, $B\mathcal{E}(r+1)$ is a finitely generated $\mathcal{E}(r+1)$ module and of course $\langle \partial H/\partial u_1, \ldots, \partial H/\partial u_r\rangle_{\mathcal{E}(r+1)}$ is finitely generated over $\mathcal{E}(r+1)$. Therefore we may apply Theorem 1.16 to deduce

(f) $$\mathcal{E}(n+r+1) = \langle \partial H/\partial x_1, \ldots, \partial H/\partial x_n\rangle_{\mathcal{E}(n+r+1)} + \langle \partial H/\partial u_1, \ldots, \partial H/\partial u_r\rangle_{\mathcal{E}(r+1)} + B\mathcal{E}(r+1).$$

(In this application of 1.16, n in Theorem 1.16 corresponds to n+r+1 here, p in 1.16 corresponds to r+1 here, f in 1.16 corresponds here to the projection $\pi\colon \mathbb{R}^n \times \mathbb{R}^{r+1} \longrightarrow \mathbb{R}^{r+1}$, C in 1.16 is $\mathcal{E}(n+r+1)$ here, A in 1.16 is $B\mathcal{E}(r+1) + \langle \partial H/\partial u_1, \ldots, \partial H/\partial u_r\rangle_{\mathcal{E}(r+1)}$ here and B in 1.16 is $\langle \partial H/\partial x_1, \ldots, \partial H/\partial x_n\rangle_{\mathcal{E}(n+r+1)}$ here.)

(Note that it is this step for which we needed, earlier on in the proof, to replace $\pi^*\mathcal{E}(1)$ by something finitely generated over $\mathbb{R}$.)

If we multiply (f) on both sides by $m(r)$, we obtain:

$$\text{(g)}\quad m(r)\mathcal{E}(n+r+1) = \langle \partial H/\partial x_1,\dots,\partial H/\partial x_n\rangle_{m(r)\mathcal{E}(n+r+1)} + \langle \partial H/\partial u_1,\dots,\partial H/\partial u_r\rangle_{m(r)\mathcal{E}(r+1)} + m(r)\mathcal{E}(r+1)B.$$

Now observe that for all $x \in \mathbb{R}^n$ and $t \in \mathbb{R}$,

$$H(x,0,t) = (1-t-s)f(x,0) + (s+t)g(x,0) = (1-s-t)\eta(x)+(s+t)\eta(x) = \eta(x).$$

So $\frac{\partial H}{\partial t}(x,0,t) = 0$, and hence $\frac{\partial H}{\partial t} \in m(r)\mathcal{E}(n+r+1)$.

Therefore, by (g), we can find germs $\xi_1,\dots,\xi_n \in m(r)\mathcal{E}(n+r+1)$, germs $\chi_1,\dots,\chi_r \in m(r)\mathcal{E}(r+1)$, and germs $\beta_0,\dots,\beta_{k+1} \in m(r)\mathcal{E}(r+1)$ (with $\beta_1=\beta_2=\dots=\beta_{k+1}=0$ in the "right" case) so that

$$\text{(h)}\quad \frac{\partial H}{\partial t}(x,u,t) = \sum_{i=1}^{n} \frac{\partial H}{\partial x_i}(x,u,t)\xi_i(x,u,t) + \sum_{j=1}^{r} \frac{\partial H}{\partial u_j}(x,u,t)\chi_j(u,t) + \sum_{\ell=0}^{k+1} \beta_\ell(u,t)H(x,u,t)^\ell .$$

Define $\omega \in \mathcal{E}(1+r+1)$ by setting $\omega(\tau,u,t) = \sum_{\ell=0}^{k+1}\beta_\ell(u,t)\tau^\ell$ for $\tau \in \mathbb{R}$, $u \in \mathbb{R}^r$, $t \in \mathbb{R}$. Equation (h) implies that the hypotheses of Lemma 1.29 are satisfied, with the following substitutions:

1.29 F n ξ η

this
application: H n+r $(\xi_1,\dots,\xi_n,\chi_1,\dots,\chi_r)$ ω

Hence by 1.29 there is a germ $\Phi = (\varphi,\psi) \in \mathcal{E}(n+r+1,n+r)$ and a germ $\lambda \in \mathcal{E}(1+n+r+1)$ such that

(i) $\Phi(x,u,0) = (x,u)$

$\lambda(\tau,x,u,0) = \tau \quad$ for $x \in R^n$, $u \in R^r$, $\tau \in R$

(j) $H(\Phi(x,u,t),t) = \lambda(H(x,u,0),x,u,t)$ for $x \in R^n$, $u \in R^r$, $t \in R$

and finally

(k) $\frac{\partial \varphi_i}{\partial t}(x,u,t) = -\xi_i(\varphi(x,u,t),\psi(x,u,t),t) \quad (i=1,\dots,n)$

$\frac{\partial \psi_j}{\partial t}(x,u,t) = -\chi_j(\psi(x,u,t),t) \quad (j=1,\dots,r)$

$\frac{\partial \lambda}{\partial t}(\tau,x,u,t) = \omega(\lambda(\tau,x,u,t),\ \psi(x,u,t),t)$

for $\tau \in R$, $x \in R^n$, $u \in R^r$, $t \in R$.

We shall show that from Φ and λ we can obtain the morphisms of unfoldings we wish to construct.

First observe that $\psi_j \in \mathcal{E}(r+1)$ for each $j (1 \leq j \leq r)$, i.e. ψ does not depend on $x \in R^n$. For $\psi(x,u,0) = u$ and hence is independent of x; and ψ satisfies the differential equations

$$\frac{\partial \psi_j}{\partial t}(x,u,t) = -\chi_j(\psi(x,u,t),t).$$

The right-hand side depends only on $\psi(x,u,t)$ but not directly on x. So by Remark 1.30, ψ does not depend on $x \in R^n$.

Similarly, λ does not depend on x, i.e. $\lambda \in \mathcal{E}(1+r+1)$.

For since $\psi \in \mathcal{E}(r+1)$ we have by (k)

$$\frac{\partial\lambda}{\partial t}(\tau,x,u,t) = \omega(\lambda(\tau,x,u,t),\ \psi(u,t),t).$$

Moreover when $t=0$, $\lambda(\tau,x,u,0) = \tau$ and does not depend on x. So again by Remark 1.30, λ does not depend on x.

Our next observation is that $\psi(0,t) = 0 \in R^r$ for all $t \in R$. For if we define $\theta \in \mathcal{E}(1,r)$ by $\theta(t) = \psi(0,t)$, then (k) implies that θ satisfies the differential equation

$$\frac{\partial\theta_j}{\partial t}(t) = -\chi_j(\theta(t),t) \qquad (j=1,\ldots,r)$$

with the initial condition $\theta(0) = 0$ (by (i)). But the constant germ $0 \in \mathcal{E}(1,r)$ also satisfies these equations; for recall that each $\chi_j \in m(r)\mathcal{E}(r+1)$, so that if we substitute 0 for $\theta(t)$ both sides of the differential equations become 0. Hence, since these differential equations have a unique solution, $\psi(0,t) = \theta(t) = 0$ for all $t \in R$.

Now this implies that $\varphi(x,0,t) = x$ for all $x \in R^n$ and $t \in R$. For by (i) this is certainly true when $t = 0$, and by (k) we have $\dfrac{\partial\varphi_i(x,0,t)}{\partial t} = -\xi_i(\varphi(x,0,t),\psi(0,t),t)$

$= -\xi_i(\varphi(x,0,t),0,t) = 0$ for all $t \in R$ and $x \in R^n$, since

$\xi_i \in m(r)\mathcal{E}(n+r+1)$.

Similarly we see that $\lambda(\tau,0,t) = \tau$ for all $\tau \in R$ and $t \in R$. For by (i) this is true when $\tau = 0$, and by (k) we have for each τ and t

$$\frac{\partial\lambda}{\partial t}(\tau,0,t) = \omega(\lambda(\tau,0,t),\psi(0,t),t)$$
$$= \omega(\lambda(\tau,0,t),0,t)\ .$$

The right-hand side is 0, by the definition of ω and since the β_ℓ are in $m(r)\mathcal{E}(r+1)$.

Note finally that in the "right" case we can find a germ $\alpha \in \mathcal{E}(r+1)$ so that $\lambda(\tau,u,t) = \tau+\alpha(u,t)$ for all $\tau \in R$, $u \in R^r$ and $t \in R$. Simply define $\alpha(u,t) = \lambda(0,u,t)$; we shall show that this α works. In the "right" case $\beta_1=\beta_2=\ldots=\beta_{k+1} = 0$; hence by the definition of ω it is clear that $\omega(\tau,u,t)$ does not depend on τ. So by (k) it follows that $\partial\lambda/\partial t$ does not depend on τ; hence $\partial^2\lambda/\partial t\partial\tau$ is identically 0 and therefore $\partial\lambda/\partial\tau$ does not depend on t. But by (i), when $t = 0$ we have $\frac{\partial\lambda}{\partial\tau}(\tau,u,0) = 1$. Hence $\partial\lambda/\partial\tau$ is everywhere 1 and so $\lambda(\tau,u,t) = \tau+\lambda(0,u,t) = \tau+\alpha(u,t)$ for all $\tau \in R$, $u \in R^r$ and $t \in R$.

Note that in fact $\alpha \in m(r)\mathcal{E}(r+1)$, since $\alpha(0,t) = \lambda(0,0,t) = 0$ for all $t \in R$.

Now, in both the "right" and the "right-left" cases, we have $\lambda(\tau,u,0) = \tau$ for all $\tau \in R$ and $u \in R^n$, so it is clear that one can find a germ $\lambda^{-1} \in \mathcal{E}(1+r+1)$ so that

$$\lambda^{-1}(\lambda(\tau,u,t),u,t) = \tau \text{ for all } \tau \in R,\ u \in R^n, \text{ and } t \in R.$$

Note that in the "right" case, if $\alpha \in \mathcal{E}(r+1)$ is chosen so that $\lambda(\tau,u,t) = \tau+\alpha(u,t)$ for all $\tau \in R$, $u \in R^r$, $t \in R$, then $\lambda^{-1}(\tau,u,t) = \tau-\alpha(u,t)$.

Now (in both the "right" and "right-left" cases again) it follows from (j) that

(l) $H(x,u,0) = \lambda^{-1}(H(\Phi(x,u,t),t),u,t)$ for $x \in R^n$, $u \in R^r$, $t \in R$.

If we first choose representatives of Φ, λ, and H, then for t near 0 we can define germs $\Phi_t \in \mathcal{E}(n+r,n+r)$ and $\lambda_t^{-1} \in \mathcal{E}(1+r)$ by setting

$$\Phi_t(x,u) = \Phi(x,u,t)$$

$$\text{and} \quad \lambda_t^{-1}(\tau,u) = \lambda^{-1}(\tau,u,t).$$

By (1) and by the definition of H we have, for t near 0,

$$F_s(x,u) = \lambda_t^{-1}(F_{s+t}(\Phi_t(x,u)),u) \qquad ((x,u) \in \mathbb{R}^{n+r}) .$$

It follows from this equation and from the previous discussion that (Φ_t, λ_t^{-1}) is a right (right-left) morphism from F_s to F_{s+t}, when t is near 0. But from (i) it is clear that for t sufficiently near 0, (Φ_t, λ_t^{-1}) is in fact a right (right-left) isomorphism, so the proof of Lemma 3.16 is complete.

§ 4 STABLE UNFOLDINGS

In this chapter we shall investigate the theory of stability of unfoldings of germs. This theory will differ from the ordinary theory of stable germs in that all of the definitions and constructions concerning stability of r-dimensional unfoldings of germs in $m(n)$ will have to respect the way R^{n+r} is fibred as $R^n \times R^r$.

The theory of stable unfoldings is of interest not only because of the importance of stability generally in the theory of singularities, but also because of its relevance to R. Thom's theory of catastrophes. In § 5, we shall apply the results of this chapter to prove the validity of Thom's celebrated list of the seven elementary catastrophes.

We shall give seven different definitions of "stable unfolding". Six of these definitions involve geometric conditions; the seventh kind of stability, infinitesimal stability, is defined by an algebraic condition. The main result of this paper (Theorem 4.11) is the proof that these definitions are all equivalent; in particular this result gives an easy-to-check algebraic criterion for the geometrically defined notions.

In this section, we shall continue to use the notational conventions of § 3.

In general, one says that something is stable if its "appearance" is not changed by small perturbations, where "appearance" is taken to mean the equivalence class with respect to a suitable equivalence relation. So we must first define the equivalence relation which will be used in the definitions of stability.

We shall model the definition of equivalence on the definition of "right-left isomorphic" in § 3. We choose "right-left isomorphic" instead of "right-isomorphic" as a model in order to make it easier for an unfolding to be stable; it is always advantageous to have as large a class of stable objects as is possible, so long as the problem of classifying the stable objects remains solvable. Moreover intuitively it seems reasonable to say that the "appearance" of an unfolding is not essentially changed by the application of diffeomorphisms on the left, so there is no reason for restricting ourselves to the "right" case.

Actually, if one did wish to make this restriction, one could develop a theory of "right-stability" analogous in all respects to the theory we shall present here. All our definitions and theorems would have analogues in this theory; the proofs would all be similar to ours and in fact somewhat easier. However, we shall not give any details of the theory of "right-stability" in this paper.

Now although our definition of equivalence will be modelled on the definition of "right-left isomorphism", it will have to differ from this definition in several important respects. For consider a germ $\eta \in m(n)$ and an r-dimensional unfolding f of η. Pick a representative f' of f on some neighbourhood U of $0 \in R^{n+r}$. In general, for a function $g': U \longrightarrow R$ which is arbitrarily close to f', the germ of $g'|\mathbb{R}^n$ will not be equal to η and in fact will not even be right-left equivalent to η unless η is non-singular. And of course we do not want to restrict our attention to unfoldings of non-singular germs. (To take a very simple example, suppose $\eta \in m(1)$ is given by $\eta(x) = x^2$, and let f be the 0-dimensional unfolding of η, i.e. $f = \eta$. (Note that η is in fact a stable germ, according to Mather's definition of stability for germs). If ε is an arbitrarily small positive real number and if g' is defined on a neighbourhood of $0 \in \mathbb{R}$ by $g'(x) = x^2+\varepsilon(x^2+x)$, then the germ of g' at 0 is non-singular, whereas η is singular at 0. So these germs cannot be right-left equivalent.)

However, what one __can__ reasonably expect is that if g' is sufficiently close to f', then for some point (x,u) in U, not necessarily the origin, the germ at (x,u) of $g'|\mathbb{R}^n\times\{u\}$, considered as a germ in m(n), will be right-left equivalent to η. (What we mean by "considered as a germ in m(n)" is: "made into a germ at the origin by prefixing a translation of $\mathbb{R}^{n+r}$, and then made into an element of m(n) by ignoring the constant term"). For instance, the example given above has this property. Note that we __cannot__ expect the germ at (x,u) of $g'|\mathbb{R}^n\times\{u\}$ to be __equal__ to η as elements of m(n), as one easily sees from the same example as before.

So from this discussion it is clear that in the definition of equivalence of unfoldings which we shall give shortly, although we will be guided by the definition of right-left

isomorphism in § 3, we shall have to omit several of the conditions which a right-left isomorphism (Φ,λ) was required to fulfill. The conditions we must omit are: (1) the requirement that Φ preserve the origin of $\mathbb{R}^{n+r}$ (2) the requirement that if $\Phi = (\varphi,\psi)$, then $\varphi|\mathbb{R}^n$ is the identity of $\mathbb{R}^n$ (or, as it is perhaps better to say since Φ need no longer preserve the origin, the requirement that $\varphi|\mathbb{R}^n$ be a translation of $\mathbb{R}^n$).
(3) the requirement that $\lambda(0,0) = 0$. (Omitting this condition will allow us to ignore the constant terms of the germs we consider at points other than the origin; i.e. it will enable us to move germs into $m(n+r)$ or into $m(n)$.)
(4) the requirement that $\lambda(\tau,0) = \tau$ for all $\tau \in \mathbb{R}$ (or, as it is again better to say, the requirement that λ_0 be a translation of $\mathbb{R}$).
We must however <u>retain</u> the requirement that the diffeomorphisms we consider respect the fibration of $\mathbb{R}^{n+r}$.

A final remark before we give the definition: Since the preceding discussion shows that we will not be able to restrict our attention to germs only at the origin of $\mathbb{R}^{n+r}$, we shall sooner or later have to begin speaking of functions defined on a neighbourhood of the origin rather than germs. It will simplify things slightly if we do this already at this stage.

These considerations lead us to the following definition:

<u>Definition 4.1.</u> Let n and r be non-negative integers. Let U and V be open subsets of $\mathbb{R}^{n+r}$ and let $f: U \longrightarrow \mathbb{R}$ and $g: V \longrightarrow \mathbb{R}$ be smooth functions. Let $(z,w) \in U$ and let $(y,v) \in V$ (where

z and y are in $\mathbb{R}^n$ and w and v are in $\mathbb{R}^r$).

We shall say that <u>f at (z,w) is equivalent to g at (y,v) as r-dimensional unfoldings</u> if for some open neighbourhood W_1 of z in $\mathbb{R}^n$ and some open neighbourhood W_2 of w in $\mathbb{R}^r$, such that $W_1 \times W_2 \subseteq U$, there exist smooth functions $\varphi: W_1 \times W_2 \longrightarrow \mathbb{R}^n$, $\psi: W_2 \longrightarrow \mathbb{R}^r$ and $\lambda: \mathbb{R} \times W_2 \longrightarrow \mathbb{R}$ satisfying the following conditions:

(a) $(\varphi(x,u), \psi(u)) \in V$ for all $x \in W_1$ and $u \in W_2$

(b) $\varphi(z,w) = y$ and $\psi(w) = v$

(c) $\varphi | W_1 \times \{w\}$ is non-singular at (z,w); ψ is non-singular at w; $\lambda | \mathbb{R} \times \{w\}$ is non-singular at $(g(y,v),w)$

(d) for all $x \in W_1$ and for all $u \in W_2$ $f(x,u) = \lambda(g(\varphi(x,u), \psi(u)), u)$

If these conditions are satisfied, we shall call the triple (φ, ψ, λ) an <u>r-dimensional equivalence</u> from f at (z,w) to g at (y,v).

Let t denote the coordinate of $\mathbb{R}$. Suppose that in addition to the above conditions, the following condition is satisfied:

(e) $\frac{\partial \lambda}{\partial t}(g(y,v),w) > 0$

In this case we shall say f at (z,w) is <u>orientedly equivalent</u> to g at (y,v), and we shall call (φ, ψ, λ) an <u>oriented equivalence.</u>

Oriented equivalence is of interest primarily for the applications to Thom's catastrophe theory, for in these applications one usually restricts one's attention to studying singular germs which have a <u>local minimum</u> at the critical point.

Actually, as we shall see, condition (e) has no effect on the stability notions we shall define. For the time being, however, we shall have to distinguish between the oriented and the non-oriented case.

It is clear that both kinds of equivalence actually define an equivalence relation on the set of ordered pairs $(f,(z,w))$, where f is a smooth function defined on an open subset U of R^{n+r} and $(z,w) \in U$.

The above definition of equivalence "induces" also a notion of equivalence for unfoldings in the following way:

Definition 4.2. Let f and g be germs in $\mathcal{E}(n+r)$. We say f and g are (orientedly) equivalent as r-dimensional unfoldings if f and g have representatives f' and g' respectively, defined on neighbourhoods of $0 \in R^{n+r}$, such that f' at 0 is (orientedly) equivalent to g' at 0.

If $\varphi \in \mathcal{E}(n+r,n)$, $\psi \in \mathcal{E}(r,r)$, and $\lambda \in \mathcal{E}(1+r)$, then (φ,ψ,λ) will be called an (oriented) equivalence from f to g if there exist representatives φ' of φ, ψ' of ψ and λ' of λ, and representatives f' of f and g' of g, such that $(\varphi',\psi',\lambda')$ is an (oriented) equivalence from f' at 0 to g' at 0.

Clearly equivalence and oriented equivalence for germs are in fact equivalence relations on $\mathcal{E}(n+r)$. When we speak of equivalence (in either sense) it will always be clear whether we are speaking of germs or of functions.

Suppose $f \in \mathcal{E}(n+r)$ unfolds $\eta \in m(n)$ and suppose $g \in \mathcal{E}(n+r)$ unfolds $\mu \in m(n)$. Observe that if f is equivalent to g, then η is right-left equivalent to μ.

For equivalence of unfoldings we have the following easy lemma.

Lemma 4.3. Suppose $f \in \mathcal{E}(n+r)$ is a right-left universal unfolding of a germ $\eta \in m(n)$. Suppose $g \in \mathcal{E}(n+r)$ unfolds $\mu \in m(n)$, and suppose f is equivalent to g (as r-dimensional unfoldings). Then g is a right-left universal unfolding of μ.

Proof: Let (φ, ψ, λ) be an equivalence from f to g. Let $\gamma = \varphi | R^n$ and let $\omega = \lambda | R$. From the definition of equivalence it is clear that $\varphi \in L(n)$ and $\omega \in L(1)$.

Now let $h \in \mathcal{E}(n+s)$ be any unfolding of μ. Define $h' \in \mathcal{E}(n+s)$ by setting $h'(x,v) = \omega(h(\gamma(x),v))$ for $x \in R^n$ and $v \in R^s$. Clearly h' unfolds η. Since f is right-left universal, there is a right-left morphism (Θ, Λ) from h' to f. Suppose $\Theta = (\theta, \chi)$ where $\theta \in \mathcal{E}(n+s,n)$ and $\chi \in \mathcal{E}(s,r)$. Define $\Theta' \in \mathcal{E}(n+s,n+r)$ and $\Lambda' \in \mathcal{E}(1+s)$ by setting $\Theta'(x,v) = (\varphi(\theta(\gamma^{-1}(x),v),\chi(v)),\psi(\chi(v)))$ and setting $\Lambda'(t,v) = \omega^{-1}(\Lambda(\lambda(t,\chi(v)),v))$ for $x \in R^n$, $v \in R^s$, $t \in R$. It is easy to check that (Θ', Λ') is a right-left morphism (of unfoldings of μ) from h to g. Since h was arbitrary, it follows that g is right-left universal.

We introduce the following notation:

Definition 4.4. Let $U \subseteq R^n$ be an open set, and let $f: U \longrightarrow R$ be a smooth function. We let $x_1, \ldots, x_n$ be the coordinates of R^n. If k is a non-negative integer and if K is a compact subset of U, we define

$\|f\|_{k,K} := \sup\{|D_\alpha f(z)| \mid z \in K;\ \alpha = (\alpha_1,\dots,\alpha_n)$ is an n-tuple of non-negative integers and $|\alpha| \leq k\}$

(Recall (Definition 1.5) that $D_\alpha f$ is the αth partial derivative of f with respect to $x_1,\dots,x_n$)

Now, before we define the stability notions we shall be investigating, recall once again that if $f \in \mathcal{E}(n+r)$ is an unfolding and if we perturb f slightly, then f will usually not be equivalent to the new unfolding at the origin, and therefore the definition of stability must involve functions defined on a neighbourhood of the origin as well as germs.

Furthermore, before we can state a definition of stability, we must make precise what we mean by a "small" perturbation. There are several ways of doing this, and each of them leads to a different notion of stability.

We begin with a rather weak notion, which was suggested by K. Jänich:

Definition 4.5. Let $\eta \in m(n)$ and let $f \in \mathcal{E}(n+r)$ be an unfolding of η. We shall say f is (orientedly) weakly stable if for every open neighbourhood U of $0 \in \mathbb{R}^{n+r}$ and for every representative f' of f defined on U, the following holds:

For every smooth function $g'\colon U \longrightarrow \mathbb{R}$ there is a real number $\varepsilon > 0$ such that for every $t \in \mathbb{R}$, if $|t| < \varepsilon$ then there is a point $(x,u) \in U$ such that $f' + tg'$ at (x,u) is (orientedly) equivalent to f' at 0.

So here we do not specify in general when a perturbation is small, but for any given perturbation we say how it can be made small enough. Essentially we have defined "small" by using an extremely fine topology on $C^\infty(U,\mathbb{R})$.

We have the following slight generalization of this definition:

Definition 4.6. Let $\eta \in m(n)$ and let $f \in \mathcal{E}(n+r)$ be an unfolding of η. We shall say f is (orientedly) weakly stable in the generalized sense if for every open neighbourhood U of $0 \in \mathbb{R}^{n+r}$ and for every representative f' of f defined on U, the following holds:

For every positive integer k and for every k-tuple $(g_1',\ldots,g_k')$ of smooth real-valued functions defined on U there exist positive real numbers $\varepsilon_1,\ldots,\varepsilon_k$ such that for all $t_1,\ldots,t_k \in \mathbb{R}$, if $|t_i| < \varepsilon_i$ $(i = 1,\ldots,k)$ then there is a point $(x,u) \in U$ such that $f'+t_1g_1'+\ldots+t_kg_k'$ at (x,u) is (orientedly) equivalent to f' at 0.

Now of course the usual and the most obvious way of defining "small" is to use the weak C^∞ topology on $C^\infty(U,\mathbb{R})$, the $\mathbb{R}$-algebra of smooth functions mapping the open set U into $\mathbb{R}$. We recall that a neighbourhood basis for this topology at $0 \in C^\infty(U,\mathbb{R})$ is given by taking all sets of the form $\{h \in C^\infty(U,\mathbb{R}) \mid \|h\|_{k,K} < \varepsilon\}$, where k varies over all non-negative integers, K varies over all compact subsets of U, and ε varies over all positive real numbers.

<u>Definition 4.7.</u> Let $\eta \in m(n)$ and let $f \in \mathcal{E}(n+r)$ be an unfolding of η. We say f is <u>(orientedly) strongly stable</u> if for every open neighbourhood U of $0 \in \mathbb{R}^{n+r}$ and every representative f' of f defined on U there is a neighbourhood V of f' in $C^\infty(U,\mathbb{R})$ (with the weak C^∞-topology) such that for every $g' \in V$ there is a point $(x,u) \in U$ such that g' at (x,u) is (orientedly) equivalent to f' at 0.

Clearly every (orientedly) strongly stable unfolding is also (orientedly) weakly stable in either sense.

Remark: It is easy to see that strong stability is a property of the equivalence classes of unfoldings $\in \mathcal{E}(n+r)$, i.e., if f and g are equivalent unfoldings, then f is strongly stable if and only if g is strongly stable. It is not at all easy to see this for weak stability. The difficulty is the following. If f and g are equivalent unfoldings and (φ,ψ,λ) is an equivalence at the origin from a representative f' of f to a representative g' of g, and if we choose all the domains of definition appropriately, then this equivalence will associate to each perturbation h_1' of g' a uniquely determined perturbation h_2' of f' such that (φ,ψ,λ) is also an equivalence at the origin from $f' + h_2'$ to $g' + h_1'$. However the mapping defined by this association will in general not be linear; in fact if h_1' is mapped into h_2' and if $t \in \mathbb{R}$, then in general th_1' will not even be mapped into a real multiple of h_2'. But it is the linear structure of $C^\infty(U,\mathbb{R})$ which plays the essential role in the definition of weak stability. This mapping will, however, be continuous with respect to the weak C^∞-topology, so there is no difficulty in the case of strong stability.

Of course the same remark remains valid if we replace equivalence by oriented equivalence and stability by oriented stability throughout.

In fact weak stability does depend only on the equivalence class of the unfolding, but this will be a consequence of our main theorem, which says that weak and strong stability are the same thing.

We wish to define one more stability concept, this time by means of an algebraic condition.

Definition 4.8. Let $\eta \in m(n)$ and let $f \in \mathcal{E}(n+r)$ be an unfolding of η. Define $F \in \mathcal{E}(n+r,1+r)$ by $F(x,u) := (f(x,u),u)$.

We say f is infinitesimally stable if

(a) $\mathcal{E}(n+r) = \langle \partial f/\partial x_1, \ldots, \partial f/\partial x_n \rangle + \langle \partial f/\partial u_1, \ldots, \partial f/\partial u_r \rangle_{\mathcal{E}(r)} + F^*\mathcal{E}(1+r)$

(Recall that we consider $\mathcal{E}(r)$ as a subring of $\mathcal{E}(n+r)$).

This definition of infinitesimal stability is motivated by the following idea. First a definition: If $h \in \mathcal{E}(p,q)$ then a smooth path germ in $\mathcal{E}(p,q)$ beginning at h is a germ $H \in \mathcal{E}(p+1,q)$ so that $H(y,0) = h(y)$ for all $y \in \mathbb{R}^p$. Given such a germ H, choose a representative H' of H and for t near $0 \in \mathbb{R}$ define $H_t \in \mathcal{E}(p,q)$ by $H_t(y) = H'(y,t)$ for $y \in \mathbb{R}^p$. Clearly the germ at 0 of the map $t \longrightarrow H_t$ does not depend on the choice of H'.

Now let $\eta \in m(n)$ and let $f \in \mathcal{E}(n+r)$ unfold η. Define germs $\iota_1 \in \mathcal{E}(n+r,n)$, $\iota_2 \in \mathcal{E}(r,r)$ and $\iota_3 \in \mathcal{E}(1+r)$ by setting $\iota_1(x,u) = x$, $\iota_2(u) = u$, and $\iota_3(s,u) = s$ for $(x,u) \in R^{n+r}$ and $s \in R$. Weak and strong stability were defined by requiring that for every sufficiently small perturbation of a representative of f there exist an equivalence between the perturbed unfolding and the representative of f. "Infinitesimal stability" should mean that this condition holds "infinitesimally" at f. That is, one could define f to be infinitesimally stable if for every germ F of a smooth path in $\mathcal{E}(n+r)$ beginning at f there exist germs φ, ψ and λ of smooth paths in $\mathcal{E}(n+r,n)$, $\mathcal{E}(r,r)$ and $\mathcal{E}(1+r)$ respectively and beginning at ι_1, ι_2 and ι_3 respectively, so that if we set $F'(x,u,t) = \lambda_t(F_t(\varphi_t(x,u),\psi_t(u)),u)$, then $\frac{\partial F'}{\partial t}(x,u,0) = 0$ for all (x,u) in R^{n+r}, or in other words, such that F' and the constant path F" (defined by $F''(x,u,t) = f(x,u)$) are "infinitesimally equal" or "tangent" at $t = 0$. But it is a simple exercise in differentiation to verify that this condition and condition 4.8(a) are equivalent, and for simplicity, we have taken 4.8(a) to be the definition of infinitesimal stability. Note, in the above motivation, that since $\lambda_o(s,u) = s$ for all $s \in R$ and $u \in R$, λ_t is "orientation-preserving" for all t near 0, so infinitesimal stability is automatically "oriented".

<u>Lemma 4.9.</u> Let $\eta \in m(n)$ and let $f \in \mathcal{E}(n+r)$ unfold η. The following conditions are equivalent:

(a) f is infinitesimally stable

(b) $$\mathcal{E}(n+r) = \langle \partial f/\partial x_1,\dots,\partial f/\partial x_n\rangle + \langle \partial f/\partial u_1,\dots,\partial f/\partial u_r\rangle_{\mathcal{E}(r)} + \mathcal{E}(r)f^*\mathcal{E}(1)$$

(c) f is right-left universal

<u>Proof:</u> We shall prove (a) ⇒ (c) ⇒ (b) ⇒ (a).

Proof that (a) ⇒ (c): Suppose f is infinitesimally stable. Define $F \in \mathcal{E}(n+r,1+r)$ by $F(x,u) := (f(x,u),u)$. Then by Definition 4.8 we have

(d) $\mathcal{E}(n+r) = \langle \partial f/\partial x_1,\dots,\partial f/\partial x_n\rangle + \langle \partial f/\partial u_1,\dots,\partial f/\partial u_r\rangle_{\mathcal{E}(r)} + F^*\mathcal{E}(1+r)$

Let $i\colon \mathbb{R}^n \longrightarrow \mathbb{R}^{n+r}$ be the canonical embedding, defined by $i(x) = (x,0)$. If we apply i^* to both sides of equation (d), we clearly obtain:

$$\mathcal{E}(n) = \langle \partial\eta/\partial x_1,\dots,\partial\eta/\partial x_n\rangle + W_f + \eta^*\mathcal{E}(1)$$

and hence by Theorem 3.22(b) f is right-left universal.

Proof of (c) ⇒ (b): Suppose f is right-left universal. Then by Theorem 3.22 we have

(e) $\mathcal{E}(n) = \langle \partial\eta/\partial x_1,\dots,\partial\eta/\partial x_n\rangle + W_f + \eta^*\mathcal{E}(1)$

Moreover, η must be finitely determined, so there is an integer k such that $m(n)^{k+1} \subseteq \langle \partial\eta/\partial x_1,\dots,\partial\eta/\partial x_n\rangle$. Now $\eta^*\mathcal{E}(1) = \langle 1,\eta,\eta^2,\dots,\eta^k\rangle_{\mathbb{R}} + \eta^*(m(1)^{k+1})$. But $\eta \in m(n)$, so $\eta^*(m(1)^{k+1}) \subseteq m(n)^{k+1} \subseteq \langle \partial\eta/\partial x_1,\dots,\partial\eta/\partial x_n\rangle$. Hence from equation (e) we get:

(f) $\mathcal{E}(n) = \langle \partial\eta/\partial x_1,\dots,\partial\eta/\partial x_n\rangle + W_f + \langle 1,\eta,\dots,\eta^k\rangle_{\mathbb{R}}$.

Now let $h \in \mathcal{E}(n+r)$. By (f) there are germs $\xi_1,\dots,\xi_n \in \mathcal{E}(n)$, real numbers $c_1,\dots,c_r$ and real numbers $b_0,\dots,b_k$ such that

(g) $h|\mathbb{R}^n = \sum_{i=1}^{n} \xi_i\,\partial\eta/\partial x_i + \sum_{j=1}^{r} c_j\alpha_j(f) + \sum_{\ell=0}^{k} b_\ell\eta^\ell$

Define $h' \in \mathcal{E}(n+r)$ by

$$h' = h - \sum_{i=1}^{n} \xi_i \partial f/\partial x_i - \sum_{j=1}^{r} c_j \partial f/\partial u_j - \sum_{\ell=0}^{k} b_\ell f^\ell$$

Then by equation (g) it follows that $h'|\mathbb{R}^n = 0$ and hence $h' \in m(r)\mathcal{E}(n+r)$ (by Lemma 1.4). Since h was an arbitrary element of $\mathcal{E}(n+r)$ we have shown:

$$\mathcal{E}(n+r) = \langle \partial f/\partial x_1, \ldots, \partial f/\partial x_n \rangle_{\mathcal{E}(n)} + \langle \partial f/\partial u_1, \ldots, \partial f/\partial u_r \rangle_{\mathbb{R}} + \langle 1, f, \ldots, f^k \rangle_{\mathbb{R}} + m(r)\mathcal{E}(n+r) .$$

And since it does no harm to make the terms on the right bigger, we have:

$$\mathcal{E}(n+r) = \langle \partial f/\partial x_1, \ldots, \partial f/\partial x_n \rangle + \langle \partial f/\partial u_1, \ldots, \partial f/\partial u_r \rangle_{\mathcal{E}(r)} + \langle 1, f, \ldots, f^k \rangle_{\mathcal{E}(r)} + m(r)\mathcal{E}(n+r) .$$

Now since $\langle \partial f/\partial u_1, \ldots, \partial f/\partial u_r \rangle_{\mathcal{E}(r)} + \langle 1, f, \ldots, f^k \rangle_{\mathcal{E}(r)}$ is a finitely generated $\mathcal{E}(r)$ module, we may apply Theorem 1.16 (Malgrange) to deduce

$$\mathcal{E}(n+r) = \langle \partial f/\partial x_1, \ldots, \partial f/\partial x_n \rangle + \langle \partial f/\partial u_1, \ldots, \partial f/\partial u_r \rangle_{\mathcal{E}(r)} + \langle 1, f, \ldots, f^k \rangle_{\mathcal{E}(r)} .$$

But since $\langle 1, f, \ldots, f^k \rangle_{\mathcal{E}(r)} \subseteq \mathcal{E}(r) f^* \mathcal{E}(1)$, it follows that (b) holds.

Proof that (b) $\Rightarrow$ (a). Define $F \in \mathcal{E}(n+r, 1+r)$ by $F(x,u) = (f(x,u),u)$. Clearly $\mathcal{E}(r) f^* \mathcal{E}(1) \subseteq F^* \mathcal{E}(1+r)$, so if (b) holds one sees immediately from Definition 4.8 that f is infinitesimally stable.

Corollary 4.10. Let $f \in \mathcal{E}(n+r)$ and $g \in \mathcal{E}(n+r)$ be r-dimensional unfoldings. Suppose f is equivalent to g. Then f is infinitesimally stable if and only if g is infinitesimally stable.

Proof: Immediate by Lemma 4.9 and Lemma 4.3.

Now we are ready to state the main result of this paper:

Theorem 4.11. Let $\eta \in m(n)$. Let $f \in \mathcal{E}(n+r)$ be an unfolding of η. The following statements are equivalent:

(a) f is infinitesimally stable

(b) f is orientedly strongly stable

(c) f is orientedly weakly stable in the generalized sense.

(d) f is orientedly weakly stable

(e) f is strongly stable

(f) f is weakly stable in the generalized sense

(g) f is weakly stable.

Proof: Clearly (b) $\Rightarrow$ (c) $\Rightarrow$ (d) $\Rightarrow$ (g) and (b) $\Rightarrow$ (e) $\Rightarrow$ (f) $\Rightarrow$ (g). So we need only prove (a) $\Rightarrow$ (b) and (g) $\Rightarrow$ (a).

Proof that (a) $\Rightarrow$ (b). Suppose f is infinitesimally stable. Let a neighbourhood U of $0 \in \mathbb{R}^{n+r}$ and a representative $f': U \longrightarrow \mathbb{R}$ of f be given. We must find a neighbourhood of f' in $C^\infty(U,\mathbb{R})$ such that any mapping in this neighbourhood is, at some point of U, orientedly equivalent (as an r-dimensional unfolding) to f' at the origin.

We begin by defining some notation: We define a function $F: U \times C^\infty(U,\mathbb{R}) \longrightarrow \mathcal{E}(n+r)$ as follows:

If $(y,v) \in U$ and $g \in C^{\infty}(U,\mathbb{R})$, define $F(y,v,g)$ by setting $F(y,v,g)(x,u) := g(y+x,v+u)-g(y,v)$ for $x \in \mathbb{R}^n$, $u \in \mathbb{R}^r$. Define also a function $\zeta\colon U \times C^{\infty}(U,\mathbb{R}) \longrightarrow m(n)$ by setting $\zeta(y,v,g) = F(y,v,g)|\mathbb{R}^n$ for all $(y,v) \in U$ and all $g \in C^{\infty}(U,\mathbb{R})$. Observe that for each non negative integer k, the functions $\pi_k \circ F\colon U \times C^{\infty}(U,\mathbb{R}) \longrightarrow J_o^k(n+r,1)$ and $\pi_k \circ \zeta\colon U \times C^{\infty}(U,\mathbb{R}) \longrightarrow J_o^k(n,1)$ are continuous. Note that $\pi_k F(y,v,g) = J_o^k g(y,v)$ and $\pi_k \zeta(y,v,g) = j_1^k g(y,v)$ for any $g \in C^{\infty}(U,\mathbb{R})$ and all $(y,v) \in U$.

Now the idea of the proof will be as follows. Suppose we are given a function $h \in C^{\infty}(U,\mathbb{R})$; we must try to construct an equivalence between f' and f'+h. As in earlier proofs, we shall consider the "homotopy" f'+th ($t \in [0,1] \subseteq \mathbb{R}$) and try to use lemma 1.29 to construct equivalences locally at each stage of the homotopy. However, in doing this we run into an important difficulty which did not occur in earlier proofs of this sort: we know that this time, the local equivalences we construct will not in general leave the origin fixed. So it will not be enough to know just that our original germ f is infinitesimally stable; to be able to apply lemma 1.29 at every stage of the homotopy we must know that there is a <u>neighbourhood</u> K of $0 \in \mathbb{R}^{n+r}$, such that for all $(y,v) \in K$, and for all $t \in [0,1]$, the germs $F(y,v,f'+th)$ are all infinitesimally stable (or equivalently right-left universal); to show that this is the case we must make a first restriction on the size of h. But there still remains a difficulty. As we go along the homotopy from time 0 to time 1, the "copies" of our original unfolding f induced by the equivalences we construct will not always be germs at 0 of f'+th, but will "move", as we have seen, along some path leading away from the origin; if this "movement" is too rapid, this path may leave the "nice" neighbourhood K before time $t = 1$; if this happens we will never reach stage $t = 1$ of the homotopy with our construction. So we must ensure that the path moves slowly and stays within K; we can do this by making further restrictions on the size of h.

One further problem remains: if we look only for equivalent copies of the original unfolding f along the homotopy, then because these copies will be found at points scattered throughout K, it will be difficult to piece all the local equivalences together to get an equivalence between f' and f'+h. Instead we must look locally for equivalent copies of all the unfoldings $F(y,v,f'+th)$, for all $(y,v) \in K$ and all $t \in [0,1]$; then we can use a compactness argument to glue everything together (K will be chosen to be compact).

So now let us begin with the proof. Recall (definition 2.21) that in Chapter 2 we defined algebraic subsets $Z_k \subset J_0^k(n,1)$ such that every jet in the complement of Z_k was right k-determining and such that conversely, if a germ in $\mathfrak{m}(n)$ was finitely determined then for large enough k its k-jet was not in Z_k.

Now f is infinitesimally stable, so by Lemma 4.9 f is right-left universal and η is finitely determined. Choose an integer k large enough so that $\pi_k(\eta) \notin Z_k$. Then since Z_k is closed in $J_o^k(n,1)$, and since $\pi_k \circ \zeta$ is continous, we have for all (y,v) in some neighbourhood of 0 in R^{n+r} that $\pi_k \zeta(y,v,f') \notin Z_k$ and therefore that $\zeta(y,v,f')$ is right k-determined. (Observe that $\zeta(0,0,f') = \eta$).

Since f is right-left universal, f is right-left k-transversal (where k is the integer we chose above). Now if g is any germ in $\mathcal{E}(n+r)$, and if g unfolds μ in $m(n)$, then by Lemma 3.13 g is a right-left k-transversal unfolding of μ if and only if

(h) $\mathcal{E}(n) = \langle \partial\mu/\partial x_1, \ldots, \partial\mu/\partial x_n \rangle + W_g + \mu^* \mathcal{E}(1) + m(n)^{k+1}$.

Now since $\mu \in m(n)$, we have $\mu^*(m(1)^{k+1}) \subset m(n)^{k+1}$. Hence equation (h) is equivalent to

(i) $\mathcal{E}(n) = \langle \partial\mu/\partial x_1, \ldots, \partial\mu/\partial x_n \rangle + W_g + \langle 1, \mu, \mu^2, \ldots, \mu^k \rangle_R + m(n)^{k+1}$

But the validity of equation (i) depends only upon the k+1-jet of g at 0. In fact, equation (i) holds if and only if $J^k(n,1)$ is generated over R by the following finite collection of elements: $\pi_k(1)$, $\pi_k(\mu), \ldots, \pi_k(\mu^k)$; $\pi_k(\alpha_1(g)), \ldots, \pi_k(\alpha_r(g))$; and the elements $\pi_k(x^\alpha \partial\mu/\partial x_i)$, where α ranges over all n-tuples of non-negative integers with $|\alpha| \leq k$ and where i ranges from 1 to n. Observe that each element of this list depends linearly upon $\pi_{k+1}(g) \in J_o^k(n+r,1)$.

Let p be the number of elements in the list above. Let q be the dimension over R of $J^k(n,1)$. (Note that $q \leq p$). Now if $g \in \mathcal{E}(n+r)$ and if g unfolds $\mu \in m(n)$, we denote the elements

$1, \mu, \ldots, \mu^k$; $\alpha_1(g), \ldots, \alpha_r(g)$; and $x^\alpha \partial\mu/\partial x_i$ $(|\alpha| \leq k,\ 1 \leq i \leq n)$

of $\mathcal{E}(n)$ in <u>some</u> order (not necessarily the order given above, but independent of the choice of g), by $\beta_1(g),\dots,\beta_p(g)$; we <u>choose</u> the ordering in such a way that $\pi_k\beta_1(f),\pi_k\beta_2(f),\dots,\pi_k\beta_q(f)$ form a basis of $J^k(n,1)$. (Recall that f is right-left k-transversal, so this is possible.)

Then clearly, since the elements $\pi_k\beta_j(g)$ depend linearly on $\pi_{k+1}(g)$, and since $\pi_{k+1}F$ is continuous on $U\times C^\infty(U,\mathbb{R})$ and $F(0,0,f') = f$, there is a neighbourhood of $0 \in \mathbb{R}^{n+r}$ such that for each (y,v) in this neighbourhood, the elements $\pi_k\beta_1(F(y,v,f')),\dots,\pi_k\beta_q(F(y,v,f'))$ form a basis of $J^k(n,1)$.

Choose a <u>compact</u> neighbourhood K of 0 in $\mathbb{R}^{n+r}$, which is small enough so that for all $(y,v) \in K$:

(j) $\pi_k\zeta(y,v,f') \notin Z_k$ and

$\pi_k\beta_1(F(y,v,f')),\dots,\pi_k\beta_q(F(y,v,f'))$ are a basis of $J^k(n,1)$.

Now choose a real number $c > 0$ which is small enough, so that the closed (n+r)-cube $[-c,c]^{n+r}$ is contained in K. We denote by $S \subset \mathbb{R}^q$ the q-1 sphere of radius c centred at the origin of $\mathbb{R}^q$.

Now we have a continuous map $\gamma: K\times S \longrightarrow J^k(n,1)$ defined by $\gamma(y,v;s_1,\dots,s_q) = \sum_{j=1}^{q} s_j\pi_k(\beta_j(F(y,v,f')))$ (for $(y,v) \in K$ and $(s_1,\dots,s_q) \in S$). Recall that $J^k(n,1)$ is isomorphic to the Euclidean space $\mathbb{R}^q$, so we have a norm $\|\ \|$ on $J^k(n,1)$ induced by the standard norm on $\mathbb{R}^q$. By the choice of K, we have that γ is never 0 on $K\times S$. Since $K\times S$ is compact there is a real number $\delta > 0$ such that for all $(y,v) \in K$ and for all $(s_1,\dots,s_q) \in S$, we have $\|\gamma(y,v;s_1,\dots,s_q)\| \geq \delta$. Since $\pi_{k+1}F$ is continuous, and since for $g \in m(n+r)$ the elements $\pi_k\beta_j(g)$ depend linearly on $\pi_{k+1}(g)$, it is clear from the compactness of K and of S that there is a real number $\varepsilon_1 > 0$ such that the following holds:

If $h \in C^\infty(U,\mathbb{R})$ and if $\|h\|_{k+1,K} < \varepsilon_1$, then for all $(y,v) \in K$ and for all $(s_1,\dots,s_q) \in S$ we have

$$\left\|\sum_{j=1}^{q} s_j\pi_k\beta_j(F(y,v,f'+h))\right\| \geq \frac{\delta}{2}$$

Observe that if $h \in C^\infty(U,\mathbb{R})$ and if $\|h\|_{k+1,K} < \varepsilon_1$, then also for any $t \in [0,1] \subseteq \mathbb{R}$ we have $\|th\|_{k+1,K} < \varepsilon_1$. Hence if $h \in C^\infty(U,\mathbb{R})$ and $\|h\|_{k+1,K} < \varepsilon_1$, we actually have

(k) $$\left\|\sum_{j=1}^{q} s_j\pi_k\beta_j(F(y,v,f'+th))\right\| \geq \frac{\delta}{2}$$

for all $(y,v) \in K$, all $(s_1,\dots,s_q) \in S$, and all $t \in [0,1] \subseteq \mathbb{R}$.

We expand our notation slightly. Define a function $\zeta': U\times C^\infty(U,\mathbb{R}) \longrightarrow \mathcal{E}(n)$ as follows: If $(y,v) \in U$ and $g \in C^\infty(U,\mathbb{R})$ define $\zeta'(y,v,g)$ by setting $\zeta'(y,v,g)(x) = g(y+x,v)$ for $x \in \mathbb{R}^n$. Note that $\zeta'(y,v,g)(x) = \zeta(y,v,g)(x) + g(y,v)$. Obviously, if ℓ is any positive integer, $\pi_\ell\circ\zeta': U\times C^\infty(U,\mathbb{R}) \longrightarrow J^\ell(n,1)$ is continuous.

Now it is clear that there is also a real number $\varepsilon_2 > 0$ such that if $h \in C^\infty(U,\mathbb{R})$ and $\|h\|_{k+1,K} < \varepsilon_2$, then

(l) $$\|\pi_k\zeta'(y,v,h)\| < \frac{\delta}{2} \quad \text{for all } (y,v) \in K .$$

Finally, recall that for $(y,v) \in K$ we have $\pi_k\zeta(y,v,f') \notin Z_k$. So since $\pi_k\zeta(y,v,g)$ for $g \in \mathcal{E}(n+r)$ depends linearly on $\pi_{k+1}g \in J^{k+1}(n+r,1)$, and since K is compact and Z_k is closed in $J^k(n,1)$, it is clear that we can choose a real number $\varepsilon_3 > 0$ such that if $h \in C^\infty(U,\mathbb{R})$ and if $\|h\|_{k+1,K} < \varepsilon_3$, then $\pi_k\zeta(y,v,f'+h) \notin Z_k$ for all $(y,v) \in K$. Again, if $\|h\|_{k+1,K} < \varepsilon_3$ then also $\|th\|_{k+1,K} < \varepsilon_3$ if $t \in [0,1]$, so we actually have: If $h \in C^\infty(U,\mathbb{R})$ and $\|h\|_{k+1,K} < \varepsilon_3$, then

(m) $\pi_k\zeta(y,v,f'+th) \notin Z_k$ for all $(y,v) \in K$ and all $t \in [0,1] \subseteq \mathbb{R}$.

Now let $\varepsilon = \min(\varepsilon_1,\varepsilon_2,\varepsilon_3)$. Let a smooth function $h\colon U \longrightarrow \mathbb{R}$ be given and suppose that $\|h\|_{k+1,K} < \varepsilon$. We shall show that there is a $(y,v) \in K \subseteq U$ such that $f'+h$ at (y,v) is orientedly equivalent to f' at 0.

Let a point $(y,v) \in K$ and a real number $a \in [0,1]$ be given. We define a germ $H \in \mathcal{E}(n+r+1)$ by setting

$H(x,u,t) = (f'+(a+t)h)(y+x,v+u)-(f'+ah)(y,v)$ for $(x,u) \in \mathbb{R}^{n+r}$ and $t \in \mathbb{R}$.

Let $\mu = H|\mathbb{R}^n$. Observe that $\mu = \zeta(y,v,f'+ah))$, so by (m) it follows that μ is right k-determined. Moreover $H|\mathbb{R}^{n+r} = F(y,v,f'+ah)$. Now from (k) it follows that the elements $\pi_k\beta_j(F(y,v,f'+ah))$, for $j = 1,\dots,q$, are linearly independent in $J^k(n,1)$ and hence are a basis of $J^k(n,1)$. This implies that $F(y,v,f'+ah)$ is a right-left k-transversal unfolding of μ, and hence is right-left universal (by theorem 3.17), since μ is right k-determined. But H unfolds $F(y,v,f'+ah)$, so H too is a right-left universal unfolding of μ. Hence by lemma 4.9 we have

$$\mathcal{E}(n+r+1) = \langle \partial H/\partial x_1,\dots,\partial H/\partial x_n\rangle + \langle \partial H/\partial u_1,\dots,\partial H/\partial u_r, \partial H/\partial t\rangle_{\mathcal{E}(r+1)}$$
$$+ \ \mathcal{E}(r+1)H^*\mathcal{E}(1).$$

If we multiply this equation by $m(r+1)$ we get

(n) $$m(r+1)\mathcal{E}(n+r+1) = \langle \partial H/\partial x_1,\dots,\partial H/\partial x_n\rangle_{m(r+1)\mathcal{E}(n+r+1)}$$
$$+ \ \langle \partial H/\partial u_1,\dots,\partial H/\partial u_r, \partial H/\partial t\rangle_{m(r+1)} + m(r+1)H^*\mathcal{E}(1).$$

Now observe that $\frac{\partial H}{\partial t}(x,u,t) = h(y+x,v+u)$ for $(x,u) \in \mathbb{R}^{n+r}$ and $t \in \mathbb{R}$. So $\frac{\partial H}{\partial t}|\mathbb{R}^n = \zeta'(y,v,h)$. Since the elements $\pi_k\beta_j(F(y,v,f'+ah))$ for $j = 1,\dots,q$ form a basis of $J^k(n,1)$, we can find an element ω of $m(n)^{k+1}$ and unique real numbers $c_1,\dots,c_q$ such that

(o) $$\zeta'(y,v,h) = \sum_{j=1}^{q} c_j\beta_j(F(y,v,f'+ah)) + \omega .$$

Moreover since $\|h\|_{k+1,K} < \varepsilon$ we have by (1) that $\|\pi_k\zeta'(y,v,h)\| < \frac{\delta}{2}$, so it follows from (k) that $|c_j| < c$ for $j = 1,\dots,q$.

Now recall that each $\beta_j(F(y,v,f'+ah))$ is one of the following elements of $\mathcal{E}(n)$:

$1,\mu,\mu^2,\dots,\mu^k$; $(\partial F(y,v,f'+ah)/\partial u_\ell)|\mathbb{R}^n$ $(1 \leq \ell \leq r)$;
or $x^\alpha\partial\mu/\partial x_i$ $(|\alpha| \leq k$ and $1 \leq i \leq n)$.

Note also that $(\partial F(y,v,f'+ah)/\partial u_\ell)|\mathbb{R}^n = (\partial H/\partial u_\ell)|\mathbb{R}^n$ since $H|\mathbb{R}^{n+r} = F(y,v,f'+ah)$.

From this and from (o) it follows that we can find germs $\xi_1,\dots,\xi_n \in \mathcal{E}(n)$, real numbers $b_1,\dots,b_r$, and a germ $\varkappa \in \mathcal{E}(1)$ such that

(p) $\frac{\partial H}{\partial t}|\mathbb{R}^n = \zeta'(y,v,h) = \sum_{i=1}^{n}\frac{\partial\mu}{\partial x_i}\xi_i + \sum_{\ell=1}^{r} b_\ell \frac{\partial H}{\partial u_\ell}|\mathbb{R}^n + \mu^*(\varkappa) + \omega$

and such that $|\xi_i(0)| < c$ $(i=1,\ldots,n)$ and $|b_\ell| < c$ $(\ell=1,\ldots,r)$.

Now $\omega \in m(n)^{k+1}$ and since μ is right k-determined it follows from Corollary 2.10 that $m(n)^{k+1} \subseteq m(n)\langle\partial\mu/\partial x_1,\ldots,\partial\mu/\partial x_n\rangle$. So choose germs $\omega_1,\ldots,\omega_n \in m(n)$ such that $\omega = \sum_{i=1}^{n}\omega_i\,\partial\mu/\partial x_i$. For $i = 1,\ldots,n$ set $\xi_i' = \xi_i+\omega_i$. Then since $\omega_i \in m(n)$ we still have $|\xi_i'(0)| < c$ $(i = 1,\ldots,n)$ and from (p) we get

(q) $\frac{\partial H}{\partial t}|\mathbb{R}^n = \sum_{i=1}^{n}\xi_i' \frac{\partial\mu}{\partial x_i} + \sum_{\ell=1}^{r} b_\ell \frac{\partial H}{\partial u_\ell}|\mathbb{R}^n + \mu^*(\varkappa)$.

Now we may consider the ξ_i' as elements also of $\mathcal{E}(n+r+1)$. If we do this we may define a germ $h' \in \mathcal{E}(n+r+1)$ by

$$h' = \frac{\partial H}{\partial t} - \sum_{i=1}^{n} \xi_i' \frac{\partial H}{\partial x_i} - \sum_{\ell=1}^{r} b_\ell \frac{\partial H}{\partial u_\ell} - H^*(\varkappa).$$

From (q) it is clear that $h'|\mathbb{R}^n = 0$, and hence $h' \in m(r+1)\mathcal{E}(n+r+1)$. So by (n) we can find germs $\gamma_1,\ldots,\gamma_n \in m(r+1)\mathcal{E}(n+r+1)$, germs $\chi_1,\ldots,\chi_{r+1} \in m(r+1)$, a germ $\rho \in m(r+1)$ and a germ $\nu \in \mathcal{E}(1)$ such that

$$h' = \sum_{i=1}^{n} \gamma_i \frac{\partial H}{\partial x_i} + \sum_{\ell=1}^{r} \chi_\ell \frac{\partial H}{\partial u_\ell} + \chi_{r+1} \frac{\partial H}{\partial t} + \rho H^*(\nu).$$

From this equation and from the definition of h' we see that

(r) $\frac{\partial H}{\partial t} - \chi_{r+1}\frac{\partial H}{\partial t} = \sum_{i=1}^{n}(\gamma_i+\xi_i')\frac{\partial H}{\partial x_i} + \sum_{\ell=1}^{r}(b_\ell+\chi_\ell)\frac{\partial H}{\partial u_\ell} + H^*(\varkappa)+\rho H^*(\nu)$

Note that since $\gamma_i \in m(r+1)$ we have $|(\gamma_i+\xi_i')(0)| = |\xi_i'(0)| < c$ for $i = 1,\ldots,n$, and since $\chi_\ell \in m(r+1)$ we have $|(b_\ell+\chi_\ell)(0)| = |b_\ell| <$ for $\ell = 1,\ldots,n$.

Now since $\chi_{r+1} \in m(r+1)$, it follows that $1-\chi_{r+1}$ is a unit in $\mathcal{E}(n+r+1)$, so we may divide both sides of equation (r) by $1-\chi_{r+1}$. Let us first simplify the notation somewhat. We set

$$\gamma_i' := (\xi_i'+\gamma_i)/(1-\chi_{r+1}) \qquad \text{(for } i = 1,\dots,n)$$

$$\chi_\ell' := (b_\ell+\chi_\ell)/(1-\chi_{r+1}) \qquad \text{(for } \ell = 1,\dots,r)$$

and we define $\theta \in \mathcal{E}(1+r+1)$ by setting

$\theta(\tau,u,t) = \kappa(\tau) + \rho(u,t)\nu(\tau)$ for $u \in \mathbb{R}^r$, $t \in \mathbb{R}$, and $\tau \in \mathbb{R}$.

Observe that $\chi_\ell' \in \mathcal{E}(r+1)$ for $\ell = 1,\dots,r$.

From (r) we get:

$$\text{(s)} \quad \frac{\partial H}{\partial t}(x,u,t) = \sum_{i=1}^n \frac{\partial H}{\partial x_i}(x,u,t)\gamma_i'(x,u,t) + \sum_{\ell=1}^r \frac{\partial H}{\partial u_\ell}(x,u,t)\chi_\ell'(u,t)$$
$$+ \theta(H(x,u,t),u,t) \quad (x \in \mathbb{R}^n;\ u \in \mathbb{R}^r;\ t \in \mathbb{R}).$$

Note also that since $(1-\chi_{r+1})(0) = 1$, we have $|\gamma_i'(0)| < c$ $(i = 1,\dots,n)$ and $|\chi_\ell'(0)| < c$ for $\ell = 1,\dots,r$.

Equation (s) implies that H satisfies the hypotheses of Lemma 1.29. From this lemma it follows that there are germs $\varphi \in \mathcal{E}(n+r+1,n)$, $\psi \in \mathcal{E}(n+r+1,r)$ and $\lambda \in \mathcal{E}(1+n+r+1)$ such that

(t) $\varphi(x,u,0) = x$

$\psi(x,u,0) = u$

$\lambda(\tau,x,u,0) = \tau$ for $x \in \mathbb{R}^n$, $u \in \mathbb{R}^r$, $\tau \in \mathbb{R}$, $t \in \mathbb{R}$

(u) $H(\varphi(x,u,t),\psi(x,u,t),t) = \lambda(H(x,u,0),x,u,t)$

$(x \in \mathbb{R}^n;\ u \in \mathbb{R}^r;\ t \in \mathbb{R})$

and such that

(v) $$\frac{\partial\varphi_i}{\partial t}(x,u,t) = -\gamma_i'(\varphi(x,u,t),\psi(x,u,t),t) \quad (i = 1,\dots,n)$$

$$\frac{\partial\psi_\ell}{\partial t}(x,u,t) = -\chi_\ell'(\psi(x,u,t),t) \quad (\ell = 1,\dots,n)$$

and $$\frac{\partial\lambda}{\partial t}(\tau,x,u,t) = \theta(\lambda(\tau,x,u,t),\psi(x,u,t),t)$$

for $x \in \mathbb{R}^n$, $u \in \mathbb{R}^r$ and $t \in \mathbb{R}$.

Since by (v) the derivatives $\partial\psi_\ell/\partial t$ do not depend directly on u, but only on $\psi(x,u,t)$ (and on t), and since $\psi(x,u,0)$ is independent of x, it follows from Remark 1.30 that ψ is independent of x, i.e. $\psi \in \mathcal{E}(r+1,r)$.

But then it follows from (v) that $\partial\lambda/\partial t$ does not depend directly on x but only on the value of λ, on u, and on t. Since $\lambda(\tau,x,u,0)$ is independent of x, it follows from 1.30 that λ does not depend on x, i.e., $\lambda \in \mathcal{E}(1+r+1)$.

Observe also, by (v), that

$$\left|\frac{\partial\varphi_i}{\partial t}(0,0,0)\right| = \left|\gamma_i'(0)\right| < c \quad (i = 1,\dots,n)$$

and $$\left|\frac{\partial\psi_\ell}{\partial t}(0,0)\right| = \left|\chi_\ell'(0)\right| < c \quad (\ell = 1,\dots,n).$$

Since $\lambda(\tau,0,0) = \tau$ for $\tau \in \mathbb{R}$, we can find a germ $\lambda^{-1} \in \mathcal{E}(1+r+1)$ such that

$$\lambda^{-1}(\lambda(\tau,u,t),u,t) = \tau$$

for all $\tau \in \mathbb{R}$, $u \in \mathbb{R}^r$ and $t \in \mathbb{R}$.

Now we may rewrite equation (u) as

(w) $$H(x,u,0) = \lambda^{-1}(H(\varphi(x,u,t),\psi(u,t),t),u,t)$$

$(x \in \mathbb{R}^n,\ u \in \mathbb{R}^r,\ t \in \mathbb{R})$.

We also observe that $\lambda^{-1}(\tau,u,0) = \tau$ for $\tau \in R$, $u \in R^r$.
In particular $\frac{\partial\lambda^{-1}}{\partial\tau}(0,0,0) = 1 > 0$.

Now choose neighbourhoods V and V' of $0 \in R^n$, neighbourhoods W and W' of $0 \in R^r$ and a neighbourhood T of 0 in R, and choose a representative of H defined on $V'\times W'\times T$, a representative of φ defined on $V\times W\times T$, a representative of ψ defined on $W\times T$ and a representative of λ defined on $R\times W\times T$. We shall use the same names for the representatives as we used for the germs(in future we shall be speaking only of the representatives). Now by making these neighbourhoods smaller if necessary we may arrange that the following holds:

(1) $\varphi(V\times W\times T) \subseteq V'$

(2) $\psi(W\times T) \subseteq W'$

(3) H is defined on $V\times W\times\{0\}$ and equation (w) holds for the representatives if $x \in V$, $u \in W$, $t \in T$.

(4) $\frac{\partial\lambda^{-1}}{\partial\tau}(H(\varphi(x,u,t),\psi(u,t),t),u,t) > 0$ for all $x \in V$, $u \in W$, $t \in T$.

(5) If $x \in V$, $u \in W$, and $t \in T$ then

$$\left|\frac{\partial\varphi_i}{\partial t}(x,u,t)\right| < c \qquad (i = 1,\dots,n)$$

and

$$\left|\frac{\partial\psi_\ell}{\partial t}(u,t)\right| < c \qquad (\ell = 1,\dots,n)$$

(6) If $x \in V'$ and $u \in W'$ then $(y+x,v+u) \in U$

(7) If $x \in V'$, $u \in W'$ and $t \in T$ then

$$H(x,u,t) = (f'+(a+t)h)(y+x,v+u)-(f'+ah)(y,v) .$$

Finally, since by (t) we have that $\varphi(x,u,0) = x$ and $\psi(u,0) = u$ when x is near enough to 0 in R^n and u is near enough to 0 in R^r, we may also arrange that

(8) If $x \in V$, $u \in W$ and $t \in T$ then

$\varphi|V\times u\times t$ is non-singular at $x \in V$

and $\psi|W\times t$ is non-singular at $u \in W$.

Now define a mapping Φ: $V\times W\times T \longrightarrow V'\times W'\times T$ by setting $\Phi(x,u,t) = (\varphi(x,u,t),\psi(u,t),t)$ for $x \in V$, $u \in W$, $t \in T$.

Condition (8) implies that Φ is non-singular everywhere on $V\times W\times T$, so it follows that $\Phi(V\times W\times T)$ is open in $V'\times W'\times T$. And clearly $0 \in \Phi(V\times W\times T)$, since by (t) we have $\Phi(0,0,0) = 0$. Choose an open neighbourhood V'' of 0 in $\mathbb{R}^n$, an open neighbourhood W'' of 0 in $\mathbb{R}^r$ and an open neighbourhood T' of 0 in $\mathbb{R}$ such that $V''\times W''\times T' \subseteq \Phi(V\times W\times T)$.

Now define a neighbourhood V_1 of $y \in \mathbb{R}^n$ and a neighbourhood W_1 of $v \in \mathbb{R}^r$ by setting $V_1 = y+V$ and $W_1 = v+W$. Similarly, define an open set $V_2(y,v,a) \subseteq \mathbb{R}^n$ and an open set $W_2(y,v,a) \subseteq \mathbb{R}^r$ by setting $V_2(y,v,a) = y+V''$ and setting $W_2(y,v,a) = v+W''$. Finally, choose a positive real number $\varepsilon(y,v,a)$ such that the open interval $(-2\varepsilon(y,v,a),2\varepsilon(y,v,a)) \subseteq T'$. And set $T_2(y,v,a) = \{t \in \mathbb{R} \mid |t-a| < \varepsilon(y,v,a)\}$.

Now for each $t \in T'$ define a smooth function φ_t: $V_1\times W_1 \longrightarrow \mathbb{R}^n$ by setting $\varphi_t(z,w) = \varphi(z-y,w-v,t)+y$, for $z \in V_1$ and $w \in W_1$; define a smooth function ψ_t: $W_1 \longrightarrow \mathbb{R}^r$ by setting $\psi_t(w) = \psi(w-v,t)+v$ for $w \in W_1$; and define a smooth function λ_t: $\mathbb{R}\times W_1 \longrightarrow \mathbb{R}$ by setting

$\lambda_t(\sigma,w) = \lambda^{-1}(\sigma-(f'+ah)(y,v),w-v,t)+(f'+ah)(y,v)$ for $\sigma \in \mathbb{R}$ and $w \in W_1$.

Now from conditions (1)-(8) it is easy to check that for each $x \in V_1$, $u \in W_1$ and for each $t \in T'$, the triple $(\varphi_t, \psi_t, \lambda_t)$ is an oriented equivalence from $f'+ah$ at (x,u) to $f'+(a+t)h$ at $(\varphi_t(x,u), \psi_t(u))$.

Moreover if $z \in V_2(y,v,a)$ and $w \in W_2(y,v,a)$ and if $t \in T'$ then there exists an $x \in V_1$ and a $u \in W_1$ such that $\varphi_t(x,u) = z$ and $\psi_t(u) = w$ (this is clear from the definition of the sets V", W", and T').

Finally, condition (5) above implies that for each $x \in V_1$, for each $u \in W_1$ and for each $t \in T'$ we have

$$\left|\frac{\partial(\varphi_t)_i}{\partial t}(x,u)\right| < c \qquad (i = 1,\dots,n)$$

and

$$\left|\frac{\partial(\psi_t)_\ell}{\partial t}(u)\right| < c \qquad (\ell = 1,\dots,r) .$$

So by these remarks, if (z,w,s) is any point of $V_2(y,v,a)\times W_2(y,v,a)\times T_2(y,v,a)$ then for any real number τ such that $|\tau| < \varepsilon(y,v,a)$, there exists a point $(z',w') \in U$ such that $f'+sh$ at (z,w) is orientedly equivalent to $f'+(s+\tau)h$ at (z',w'), and such that $|z_i-z_i'| < |\tau|c$ $(i = 1,\dots,n)$ and $|w_\ell-w_\ell'| < |\tau|c$ $(\ell = 1,\dots,r)$. For by the remarks above, since $s-a \in T'$, there is a point $(x,u) \in V_1\times W_1$ such that $f'+sh$ at (z,w) is orientedly equivalent to $f'+ah$ at (x,u); but if $s \in T_2(y,v,a)$ and if $|\tau| < \varepsilon(y,v,a)$ then $s+\tau-a \in T'$, so there is a point $(z',w') \in U$ such that $f'+ah$ at (x,u) is orientedly equivalent to $f'+(s+\tau)h$ at (z',w'). Hence $f'+sh$ at (z,w) is orientedly equivalent to $f'+(s+\tau)h$ at (z',w').

Moreover we can choose (x,u) and (z',w') such that $(z,w) = (\varphi_{s-a}(x,u),\psi_{s-a}(u))$ and $(z',w') = (\varphi_{s-a+\tau}(x,u),\psi_{s-a+\tau}(u))$.

But if t is between $s-a$ and $s-a+\tau$, then $t \in T'$, and for all $t \in T'$ we have $\left|\frac{\partial(\varphi_t)_i}{\partial t}(x,u)\right| < c$ $(i = 1,\ldots,n)$ and $\left|\frac{\partial(\psi_t)_\ell}{\partial t}(x,u)\right| < c$ $(\ell = 1,\ldots,r)$. Hence clearly $|z_i-z_i'| < |\tau|c$ $(i = 1,\ldots,n)$ and $|w_\ell-w_\ell'| < |\tau|c$ $(\ell = 1,\ldots,r)$.

Now $K\times[0,1]$ is compact, and hence can be covered by finitely many of the sets $V_2(y,v,a)\times W_2(y,v,a)\times T_2(y,v,a)$ (for $(y,v) \in K$ and $a \in [0,1]$). To each of these finitely many sets corresponds a positive real number $\varepsilon(y,v,a)$; let ε' be the minimum of these finitely many numbers.

Then for any $(y,v) \in K$ and any $a \in [0,1]$, if $t \in \mathbb{R}$ and $|t| < \varepsilon'$, then there is a $(z,w) \in U$ such that $f'+ah$ at (y,v) is orientedly equivalent to $f'+(a+t)h$ at (z,w) and such that $|z_i-y_i| < |t|c$ $(i = 1,\ldots,n)$ and $|w_\ell-v_\ell| < |t|c$ $(\ell = 1,\ldots,r)$.

Choose a positive integer N large enough so that $1/N < \varepsilon'$. From the preceding paragraph it is clear that, by induction, one can find points $(z^{(j)},w^{(j)}) \in K$, for $j = 0,\ldots,N$, such that:

f' at 0 is orientedly equivalent to $f'+(j/N)h$ at $(z^{(j)},w^{(j)})$

$|z_i^{(j)}| \leq jc/N$ $(i = 1,\ldots,n)$

and

$|w_\ell^{(j)}| \leq jc/N$ $(\ell = 1,\ldots,r)$.

The estimate on $|z_i^{(j)}|$ and $|w_\ell^{(j)}|$ ensures that the induction can be continued up to $j = N$, since c was chosen small enough so that $[-c,c]^{n+r} \subseteq K$.

Now $(z^{(N)},w^{(N)}) \in K \subseteq U$, and f' at 0 is orientedly equivalent to f'+h at $(z^{(N)},w^{(N)})$, so this completes the proof.

Proof that (g) $\Rightarrow$ (a): Suppose $f \in \mathcal{E}(n+r)$ unfolds $\eta \in m(n)$, and suppose f is weakly stable. We must show that f is infinitesimally stable.

The idea of the proof is to use the Thom transversality lemma to find a function, defined on a neighbourhood of the origin, which is close enough to a representative f' of f to be equivalent at some point to f' at 0, and whose germs <u>everywhere</u> on its domain of definition are infinitesimally stable.

We recall that a germ $h \in \mathcal{E}(n+r)$, which unfolds $\mu \in m(r)$, is an infinitesimally stable unfolding of μ if and only if it is right-left universal, and this is the case if and only if h is right-left k-transversal for every non-negative integer k. Recall also that h is right-left k-transversal if and only if $j_1{}^k h$ is transversal at 0 to the orbit $L^k(1)\pi_k(\mu)L^k(n)$.

Now suppose g is some smooth real valued function defined on an open neighbourhood U of $0 \in R^{n+r}$. We have the standard imbedding $i: R^n \longrightarrow R^{n+r}$ defined by $i(x) = (x,0)$ for $x \in R^n$. We recall that i induces a linear map ${}_k i^*: J_o^k(n+r,1) \longrightarrow J_o^k(n,1)$ for each k $(0 \leq k < \infty)$; clearly ${}_k i^*$ is just a projection of Euclidean spaces.

We also have for each k a projection $\rho_k: J^k(n+r,1) \longrightarrow J_o^k(n+r,1)$; ρ_k just "forgets" the constant term of a jet. Recall from the definition of the function $j_1^k g: U \longrightarrow J_o^k(n,1)$ that

$j_1^k g = {}_k i^* J_o^k g = {}_k i^* \rho_k J^k g.$

If k is a non-negative integer, we set
$A_k := \rho_k^{-1}({}_k i^*)^{-1}(L^k(1)\pi_k(\eta)L^k(n))$. Clearly A_k is a submanifold of $J^k(n+r,1)$. Moreover, since ${}_k i^*$ and ρ_k are projections, it is clear that if $(x,u) \in U$, then $j_1^k g$ is transversal at (x,u) to $L^k(1)\pi_k(\eta)L^k(n)$ if and only if $J^k g$ is transversal at (x,u) to A_k.

Choose a neighbourhood U of the origin in $\mathbb{R}^{n+r}$ and a representative f' of f defined on U. We shall show that there is a function $h \in C^\infty(U,\mathbb{R})$, such that there exist real numbers t arbitrarily close to 0 such that for every non-negative integer k, $J^k(f'+th)$ is transversal to A_k everywhere on U.

If k is a non-negative integer, and if $t \in \mathbb{R}$ and $t \neq 0$, let $V_t^k := \{h \in C^\infty(U,\mathbb{R}) \mid J^k(f'+th)$ is transversal to A_k everywhere on $U\}$. Then it follows trivially from the Thom transversality lemma (Corollary 1.23) that for each k $(0 \leq k < \infty)$, and for each non-zero $t \in \mathbb{R}$, V_t^k is a countable intersection of open dense subsets of $C^\infty(U,\mathbb{R})$. Now let $V := \bigcap_{k=0}^\infty \bigcap_{m=1}^\infty V_{1/m}^k$. V is also a countable intersection of open dense subsets of $C^\infty(U,\mathbb{R})$, so since $C^\infty(U,\mathbb{R})$ is a Baire space, V is dense, and in particular non-empty. Pick an $h \in V$.

Now f is weakly stable, so there is a real number $\varepsilon > 0$, such that if $t \in \mathbb{R}$ and $|t| < \varepsilon$, then f' at 0 is equivalent to f'+th at some point of U. Since $h \in V$, we can find a real number s such that $|s| < \varepsilon$ and such that for each k $(0 \leq k < \infty)$, $J^k(f'+sh)$ is transversal to A_k everywhere on U. Pick a suitable point $(y,v) \in U$ and an equivalence (φ,ψ,λ) from f'+sh at (y,v) to f' at 0.

For each k ($0 \leq k < \infty$), $J^k(f'+sh)$ is transversal to A_k at (y,v), so $j_1^k(f'+sh)$ is transversal to $L^k(1)\pi_k(\eta)L^k(n)$ at (y,v). Define a germ $g \in \mathcal{E}(n+r)$ by setting $g(x,u) = (f'+sh)(y+x,v+u) -(f'+sh)(y,v)$ for $(x,u) \in R^{n+r}$; clearly $j_1^k g$ is transversal to $L^k(1)\pi_k(\eta)L^k(n)$ at 0 for every non-negative integer k.

Now define germs $\varphi' \in \mathcal{E}(n+r,n)$, $\psi' \in \mathcal{E}(r,r)$ and $\lambda' \in \mathcal{E}(1+r)$ by setting $\varphi'(x,u) = \varphi(y+x,v+u)$; $\psi'(u) = \psi(v+u)$; and $\lambda'(t,u) = \lambda(t,v+u)-(f'+sh)(y,v)$, for $x \in R^n$, $u \in R^r$ and $t \in U$.

Obviously $(\varphi',\psi',\lambda')$ is an equivalence from g to f. Set $\mu := g|R^n$. Since g is equivalent to f, it follows that μ is right-left equivalent to η, so for every non-negative integer k, we have $L^k(1)\pi_k(\mu)L^k(n) = L^k(1)\pi_k(\eta)L^k(n)$.

Hence for every k ($0 \leq k < \infty$), $j_1^k g$ is in fact transversal at 0 to $L^k(1)\pi_k(\mu)L^k(n)$, and g is therefore right-left k-transversal for all k. By Lemma 3.18, this implies g is right-left universal and hence infinitesimally stable.

Since f and g are equivalent, f too is infinitesimally stable, by Corollary 4.10. This completes the proof of theorem 4.11.

The most important consequence of theorem 4.11 is, of course, that we may henceforth speak simply of "stable unfoldings" without having to worry about what kind of stability is meant. An equally important corollary is that we have easily verifiable algebraic criteria for stability: namely, condition 4.8(a), which was used to define infinitesimal stability, or equivalently, the criteria for right-left universality which were given in Chapter 3.

§ 5 THE SEVEN ELEMENTARY CATASTROPHES: A CLASSIFICATION THEOREM

In this section we shall prove the validity of Rene Thom's celebrated list of the "seven elementary catastrophes".

In Thom's theory of catastrophes, the mathematical model for a natural process is obtained in many cases by considering the set of minima of a smooth family, indexed by a manifold C, of potential functions defined on some manifold M. The manifold C, called the "control space", is the space in which the process is observed or in which it takes place, and is usually 4-dimensional space-time. M, the "state space", can be of high dimension and is parametrized by the various states which are involved in a physical description of the process. Locally, near a minimum, such a family of potential functions is of course just an unfolding of a singular germ.

Thom has claimed that there are only seven essentially different unfoldings which can be involved in the local description of a natural process of the above type; the catastrophes defined by these unfoldings he calls the seven elementary catastrophes. The list of the seven elementary catastrophes has achieved wide distribution; but it is not well known among mathematicians what the list means exactly, or why there are only these seven elementary catastrophes. It will be our goal in this section to suggest a precise interpretation of Thom's claim and to supply the proof.

A more detailed discussion of Thom's theory of catastrophes and its relation to the material of this chapter will be found in the appendix. Here we shall restrict ourselves to presenting the relevant mathematical results.

We continue to use the notational conventions of § 3. We begin with a definition which will enable us to present Thom's list in a concise form:

Definition 5.1. Let $\mu \in m(n)$ and let $g \in \mathcal{E}(n+r)$ be an unfolding of μ. Let $\eta \in m(n+q)$ and let $f \in \mathcal{E}(n+q+r+s)$ unfold η. Let λ be an integer, $0 \leq \lambda \leq q$. We say f reduces orientedly to g with index λ if f is orientedly equivalent (as an r+s-dimensional unfolding) to the unfolding $g' \in \mathcal{E}(n+q+r+s)$ given by

(a) $g'(x,y,u,v) = g(x,u)-y_1^2-y_2^2-\dots-y_\lambda^2+y_{\lambda+1}^2+\dots+y_q^2$

$$(x \in \mathbb{R}^n,\ y \in \mathbb{R}^q,\ u \in \mathbb{R}^r,\ v \in \mathbb{R}^s)$$

We shall say f reduces to g with index λ if f reduces orientedly with index λ to g or to -g.

If f reduces (orientedly) to g with index λ, we call g an (oriented) reduction of f of index λ.

The index of a reduction will not always play a role in the discussion. We shall say simply f reduces (orientedly) to g if for some λ $(0 \leq \lambda \leq q)$, the unfolding f reduces (orientedly) to g with index λ.

If, above, q+s is positive (i.e. non-zero) we shall say f reduces (orientedly and) properly to g, and g will be called a proper (oriented) reduction of f.

If an unfolding f has no proper reduction, then f will be said to be irreducible. (Note that if f has no proper oriented reduction, then clearly f also has no proper (unoriented) reduction, so there is no need to define a separate notion of oriented irreducibility).

Remarks: We observe that if f and g are as above, then it is easy to see that f reduces to g if and only if there is an unfolding g' of the form 5.1(a) (for some λ) such that f is equivalent (not necessarily orientedly) to g'. The reason we did not define (unoriented) reduction in this simpler way is to make the index of a reduction as meaningful as possible. In general the index of a reduction is not uniquely determined. For example let $n = 2$, $q = 1$, $r = s = 0$, and let $\mu(x,y) = x^3-y^2$ and $\eta(x,y,z) = x^3-y^2+z^2$ $(x,y,z \in R)$. Since $r = s = 0$ we have $f = \eta$ and $g = \mu$. Clearly f reduces to g with index 0. However f also reduces to g with index 1. For if we define $\varphi(x,y,z) = (-x,z,y)$, then f is orientedly equivalent to $f\varphi$ and $f\varphi(x,y,z) = -x^3+y^2-z^2 = -g(x,y)-z^2$. For oriented reductions, however, the index is uniquely determined (if η and μ are singular) because if f reduces orientedly to g with index λ, then λ equals the index of the Hessian form of η minus the index of the Hessian form of μ. But because we defined (unoriented) reduction in the way we did, it will also turn out (as a consequence (Corollary 5.14) of the "splitting lemma" 5.12) that if g is stable and is an irreducible reduction of f then the index is uniquely determined, even if we do not consider only oriented reductions (again assuming η and μ are singular).

For it will follow from the splitting lemma that if g is irreducible, then the Hessian form of μ is 0 and hence μ and $-\mu$ have the same Hessians.

This is of importance for Thom's list, since all of the unfoldings which occur in the list are irreducible.

<u>More remarks:</u> Note that if f, g, and h are unfoldings, and if f reduces to g and g reduces to h, then f reduces to h. In fact, the relation "reduces to" induces a partial ordering on the set of equivalence classes of unfoldings.

Obviously every unfolding has an irreducible reduction (just take a reduction in $\mathcal{E}(p)$ for minimal p).

Of course, these remarks are also valid in the oriented case.

The reason we consider reductions is the following:

If $f \in \mathcal{E}(n+r)$ unfolds $\eta \in \mathfrak{m}(n)$, pick a representative f' of f and define the <u>singular locus</u> of f to be the germ at 0 of the subset of $\mathbb{R}^{n+r}$ consisting of all singular points of the functions $f'|\mathbb{R}^n \times u$ for u near 0 in $\mathbb{R}^r$. Clearly this definition is independent of the choice of f'. Now of course it is the singular loci of unfoldings which are of primary interest to the theory of singularities, not the unfoldings themselves. But taking a reduction of an unfolding does not essentially change the singular locus. For if f reduces to g, then f can be obtained (up to equivalence) from g by adding to g a non-degenerate quadratic form in new variables and then taking a constant unfolding. But adding a non-degenerate quadratic form just raises the dimension of the Euclidean space in which the singular locus is embedded; it does not change the singular locus itself. And taking a constant unfolding just suspends the singular locus, i.e., replaces it by its Cartesian product with a Euclidean space. So by reducing an unfolding, we just bring it into a simpler form without affecting its structure in an essential way.

For the Thom theory, one needs to know somewhat more than just the singular locus of an unfolding f. One needs to know also which of the points in the singular locus are local minima of the restrictions to the $\mathbb{R}^n$-fibres of representatives of f, and which are not. This is why the index of a reduction is important to the Thom theory. For although the locus of minima is not invariant under reductions in general, it does remain essentially the same (in the same sense as above) under oriented reductions of index 0, as is clear by the same argument as above.

We can make a similar definition for germs:

Definition 5.2. Let $\mu \in m(n)$ and let $\eta \in m(n+q)$. Let λ be an integer, $0 \leq \lambda \leq q$. We say η reduces on the right to μ with index λ if η is right-equivalent to the germ $\mu' \in m(n+q)$ given by

(a) $\mu'(x,y) = \mu(x)-y_1^2-\dots-y_\lambda^2+y_{\lambda+1}^2+\dots+y_q^2 \quad (x \in \mathbb{R}^n, y \in \mathbb{R}^q)$.

We shall say η reduces on both sides to μ with index λ if there is a germ $\omega \in L(1)$ such that $\partial\omega/\partial t(0) > 0$ and such that $\omega\eta$ reduces on the right with index λ to $\pm\mu$.

If η reduces on the right (on both sides) to μ with index λ, we call μ a right (right-left) reduction of η of index λ.

Again, the index of a reduction will not always play a role in the discussion. We shall say simply η reduces on the right (on both sides) to μ if for some λ $(0 \leq \lambda \leq q)$, the germ η reduces on the right (on both sides) to μ with index λ.

If $q > 0$, we say η reduces properly on the right (on both sides) to μ and we call μ a proper right (right-left) reduction of η.

We observe that a germ η has proper right reductions if and only if η has proper right-left reductions. "Only if" is trivial; "if" will be clear from the following observations:

If q is a non-negative integer and $0 \leq \lambda \leq q$, define $Q_\lambda \in m(q)$ by:

$$Q_\lambda(y) = -y_1^2-\dots-y_\lambda^2+y_{\lambda+1}^2+\dots+y_q^2 \quad (y \in R^q) .$$

Now suppose ν is a germ in $m(n)$ for some n, and suppose $\omega \in L(1)$; define $\nu' \in m(n+q)$ by

$$\nu'(x,y) = \omega(\nu(x)+Q_\lambda(y)) \text{ for } x \in R^n,\ y \in R^q.$$

We claim that there is a germ $\alpha \in m(n)$ and an integer λ' $(0 \leq \lambda' \leq q)$, such that if we define $\alpha' \in m(n+q)$ by: $\alpha'(x,y) = \alpha(x)+Q_{\lambda'}(y)$ $(x \in R^n,\ y \in R^q)$, then $\nu' \sim_r \alpha'$.

To show this we observe that for any $y \in R^q$ we have $\nu'(0,y) = \omega Q_\lambda(y)$; hence we may consider ν' as an n-dimensional unfolding of the germ $\omega Q_\lambda \in m(q)$. One computes immediately that r-codim$(Q_\lambda) = 0$; hence also r-codim$(\omega Q_\lambda) = 0$ and so ωQ_λ is a right-universal unfolding of itself of minimal unfolding dimension. It follows (by 3.21) that the unfolding ν' is right-isomorphic to the n-dimensional constant unfolding of ωQ_λ, i.e., there is a $\varphi \in L(n+q)$ and a germ $\alpha \in m(n)$ such that $\nu'\varphi(x,y) = \omega Q_\lambda(y) + \alpha(x)$. Finally, we note that since Q_λ is a non-degenerate quadratic form on R^q, and since $\omega \in L(1)$, it follows that ωQ_λ is a non-degenerate quadratic form on R^q and hence, by the Morse Lemma, that $\omega Q_\lambda \sim_r Q_{\lambda'}$, for some λ'. Combining these facts, we see that $\nu' \sim_r \nu'\varphi = \omega Q_\lambda + \alpha \sim_r Q'_\lambda + \alpha = \alpha'$. This proves the claim above.

Now if $\eta \in m(p)$ has a proper right-left reduction, then there is an integer q $(0 < q \leqslant p)$, an integer λ $(0 \leqslant \lambda \leqslant q)$, a germ $\nu \in m(p-q)$ and a germ $\omega \in L(1)$ such that $\omega\eta \sim_r \nu + Q_\lambda$. Hence $\eta \sim_r \omega^{-1}(\nu + Q_\lambda)$, and by our observation above, there is a λ' $(0 \leqslant \lambda' \leqslant q)$ and a germ $\alpha \in m(p-q)$ such that $\omega^{-1}(\nu + Q_\lambda) \sim_r \alpha + Q'_\lambda$. Hence η reduces properly on the right to α, which is what we wished to show. This completes the proof of "if".

If a germ η has no proper right reductions (or equivalently, no proper right-left reductions) we shall say η is <u>irreducible</u>.

As for unfoldings, so for germs it is clear that every germ has irreducible right (right-left) reductions.

We do not bother to give the obvious definition of oriented reduction of germs, since we will not need this concept.

Remark: If η and μ are as in Definition 5.2., then clearly η reduces on both sides to μ if and only if there is a germ μ' of the form 5.2 (a) (for some λ), such that $\eta \sim_{rl} \mu'$. Here again (as was the case for unfoldings) the reason for not defining right-left reduction by means of this slightly simpler condition is to ensure that the index of a right-left reduction will be as meaningful as possible. Again it will follow from Corollary 5.14 that the index of an irreducible right-left reduction of a singular germ η is uniquely determined, although in general the index is not uniquely determined, as our previous example shows.

Remark: One can show that a germ $\eta \in m(n+q)$ reduces on both sides to a germ $\mu \in m(n)$ if and only if η reduces on the right to a germ in $m(n)$ which is right-left equivalent to μ. However, the proof of this fact is quite difficult and long, so we do not give it here in full. The basic idea is as follows:

If η reduces on both sides to μ, then for some λ $(0 \leqslant \lambda \leqslant q)$ and for some $\omega \in L(1)$, we have $\eta \sim_r \omega\mu'$, where μ' is given by 5.2 (a), i.e. $\mu'(x,y) = \mu(x)+Q_\lambda(y)$ $(x \in R^n,\ y \in R^q)$ (here $Q_\lambda \in m(q)$ is defined as on the previous page). Now one can show that $\omega(\mu+Q_\lambda) \sim_r \omega\mu+\omega Q_\lambda$. This is so because $\omega(\mu+Q_\lambda)$ and $\omega\mu+\omega Q_\lambda$ are both right-universal n-dimensional unfoldings of ωQ_λ and are therefore right-isomorphic. At first sight, this implies only that for some $\alpha \in m(n)$ we have $\omega(\mu+Q_\lambda) \sim_r \omega\mu+\omega Q_\lambda+\alpha$; however, by mimicking the proof of Lemma 3.16 one can show that in this particular case α can in fact be taken to be 0; thus we have $\omega(\mu+Q_\lambda) \sim_r \omega\mu+\omega Q_\lambda$. By the Morse lemma, $\omega Q_\lambda \sim_r Q'_\lambda$ for some λ'. Combining these facts, we find that $\eta \sim_r \omega\mu+Q'_\lambda$, so η reduces on the right to $\omega\mu$. The converse is proved similarly.

A final remark: If f unfolds η and if g unfolds μ, and if f reduces to g with index λ, then clearly η reduces to μ (on both sides) with index λ.

<u>Definition 5.3.</u> Let R be the smallest equivalence relation on the set of all unfoldings such that if f and g are unfoldings and f reduces to g, then fRg.

If f and g are unfoldings, we say f is <u>similar</u> to g if and only if fRg.

Let R' be the smallest equivalence relation on $\bigcup_{n=1}^{\infty} m(n)$ such that if $\eta \in m(n)$ and $\mu \in m(n')$ and η reduces on both sides to μ, then $\eta R'\mu$.

If $\eta \in m(n)$ and $\mu \in m(n')$, we say η is <u>similar</u> to μ if and only if $\eta R'\mu$.

As we have pointed out in a previous remark, similar unfoldings (and similar germs) have essentially the same structure, as far as the theory of singularities is concerned.

Observe that if f unfolds η and g unfolds μ, and if f is similar to g, then η is similar to μ.
The following definition will enable us to eliminate an uninteresting case when we state Thom's list.

<u>Definition 5.4.</u> Let $f \in \mathcal{E}(n+r)$ unfold $\eta \in m(n)$. Let λ be an integer $(0 \leqslant \lambda \leqslant n)$. We say f has a <u>simple singularity of index λ</u> at 0 if η is right-equivalent to the germ $Q_\lambda \in m(n)$ defined by

$$Q_\lambda(x_1,\ldots,x_n) = -x_1^2-x_2^2-\ldots-x_\lambda^2+x_{\lambda+1}^2+\ldots+x_n^2 \quad \text{for } (x_1,\ldots,x_n) \in R^n$$

If $\lambda = 0$, we say f has a <u>simple minimum</u> at 0. If $\lambda = n$, we say f has a <u>simple maximum</u> at 0.

<u>Remarks:</u> Observe that here λ is uniquely determined, since by the Morse lemma f has a simple singularity of index λ at 0 if and only if η has a non-degenerate critical point of index λ at 0, and the index of a non-degenerate critical point is well-defined.

If f has a simple singularity at 0, then f is stable. In fact, f is right-universal. For clearly $\langle \partial Q_\lambda/\partial x_1, \ldots, \partial Q_\lambda/\partial x_n\rangle = \langle x_1, \ldots, x_n\rangle = m(n)$, so r-codim $Q_\lambda = 0$. This implies that r-codim$(\eta) = 0$, so 0 is the minimal dimension of a right-universal unfolding of η. This means η is a right universal unfolding of itself, so since f unfolds η, f is right universal.

Moreover it then follows that f is right-isomorphic to a constant unfolding of η. This means that there is no Thom catastrophe associated to f, so we must exclude this case when we present Thom's list. (For a definition of "catastrophe", see the appendix).

From the preceding discussion it is clear that if f has a simple singularity at 0, then f reduces orientedly to the trivial unfolding $0 \in m(0)$.

The converse also holds: If $f \in \mathcal{E}(n+r)$ unfolds $\eta \in m(n)$ and f is similar to the unfolding $0 \in m(0)$, then f has a simple singularity at 0. Moreover if η has a local minimum (maximum) at 0, then f has a simple minimum (maximum) at 0. For if f is similar to $0 \in m(0)$ then η is similar to $0 \in m(0)$; it follows that η has a non-degenerate critical point at 0, for this property is clearly invariant under similarity of germs. Moreover if η has a local minimum (maximum) at 0, the index of the critical point must clearly be 0 (respectively n).

Finally, we need one more definition before we can give Thom's list:

Definition 5.5. Let $f \in \mathcal{E}(n+r)$ unfold $\eta \in m(n)$. We say f has local minima near 0 if for every neighbourhood U of $0 \in \mathbb{R}^{n+r}$ and for every representative f' of f defined on U, there is a point $(x,u) \in U$ such that the function $f' | ((\mathbb{R}^n \times \{u\}) \cap U)$ has a local minimum at (x,u).

Note that in particular if η has a local minimum at 0, then f has local minima near 0. However f can have local minima near 0 even if η does not have a minimum at 0. This is the case, for example, for $\eta(x) = x^3$ and $f(x,u) = x^3+ux$.

Remark: We do not use "local minimum" in the strict sense. If g is a real-valued function defined near a point y, we say g has a local minimum at y if $g(y) \leq g(z)$ for z near y.

We are now ready to state the main theorem of this section:

Theorem 5.6. (Thom's "list of the seven elementary catastrophes") Let $f \in \mathcal{E}(n+r)$ be an unfolding of $\eta \in m(n)^2$. Suppose f is stable and has local minima near 0, and suppose $r \leq 4$. Then either f has a simple minimum at 0, or f reduces with index 0 to one of the following seven irreducible unfoldings g_i of germs μ_i:

Name	μ_i	g_i	unfolding dimension
fold	$\mu_1(x)=x^3$	$g_1(x,u)=x^3+ux$	1
cusp	$\mu_2(x)=x^4$	$g_2(x,u,v)=x^4+ux^2+vx$	2
swallowtail	$\mu_3(x)=x^5$	$g_3(x,u,v,w)=x^5+ux^3+vx^2+wx$	3
butterfly	$\mu_4(x)=x^6$	$g_4(x,u,v,w,t)=x^6+ux^4+vx^3+wx^2+tx$	4
hyperbolic umbilic (wave crest)	$\mu_5(x,y)=x^3+y^3$	$g_5(x,y,u,v,w)=x^3+y^3+uxy+vx+wy$	3
elliptic umbilic (hair)	$\mu_6(x,y)=x^3-xy^2$	$g_6(x,y,u,v,w)=x^3-xy^2+u(x^2+y^2)+vx+wy$	3
parabolic umbilic (mushroom)	$\mu_7(x,y)=x^2y+y^4$	$g_7(x,y,u,v,w,t)=x^2y+y^4+ux^2+vy^2+wx+ty$	4

Moreover, no two of the unfoldings in this list are similar to each other, nor are any of them similar to the trivial unfolding $0 \in m(0)$.

Note: The hypothesis that f have local minima near 0 plays no role in the classification; it merely ensures that the reductions have index 0. If f does not have local minima near 0, but fulfills the other hypotheses, then either f has a simple singularity (of some index) at 0, or f reduces (but not with index 0) to one of the seven unfoldings g_i. This will be clear from the proof.

Note that each of the unfoldings g_i in the list is a minimal right-universal and also a minimal right-left universal unfolding of μ_i, as is easily checked using Theorem 3.22, for example.

One also easily sees that each of the seven unfoldings g_i has local minima near 0. In fact, if μ is a germ in $m(p)^3$ for some p, then any stable unfolding g of μ has local minima near 0, for g right-left induces the unfolding $h \in \mathcal{E}(p+1)$ given by $h(x,u) = \mu(x)+u\Sigma_{i=1}^{p}x_i^2$, and since $\mu \in m(p)^3$ and $\Sigma_{i=1}^{p}x_i^2$ is right 2-determined, it follows that h has a local minimum (on the fibre) at $(0,u)$ for all $u > 0$, and a local maximum at $(0,u)$ for all $u < 0$. Hence g has local minima near 0, clearly.

From these remarks it follows that each of the unfoldings in the list (and also the trivial unfolding $0 \in m(0)$) satisfies the hypotheses of the theorem, i.e., all of these unfoldings can actually occur in the classification.

Finally observe that since each g_i is stable and each of the germs μ_i has vanishing two-jet, it follows (by an argument involving the Hessian which was sketched in the remark after definition 5.1) that if an unfolding f reduces to one of the g_i, then the index of the reduction is uniquely determined (See Corollary 5.14).

From Theorem 5.6 one easily obtains a classification of unfoldings by means of <u>oriented</u> reductions. For by definition, if an unfoldin f reduces with index 0 to one of the unfoldings g_i listed in Theorem 5.6, then f reduces orientedly with index 0 to $\pm g_i$; moreover unless i = 2,4 or 7, $-g_i$ and g_i are orientedly equivalent, as is easy to see. So all one need do to obtain a classification by oriented reductions is to add $-g_2$, $-g_4$ and $-g_7$ to the list of theorem 5.6. We give the expanded list explicitly below as Theorem 5.7; in this list we have replaced

$-g_2, -g_4$ and $-g_7$ by unfoldings of a more aesthetic form which are obviously orientedly equivalent to them.

<u>Theorem 5.7.</u> Let $f \in \mathcal{E}(n+r)$ be an unfolding of a germ $\eta \in m(n)^2$. Suppose f is stable and has local minima near 0, and suppose $r \leq 4$. Then either f has a simple minimum at 0, or f reduces orientedly with index 0 to one of the following 10 irreducible unfoldings g_i of germs μ_i

Name	μ_i	g_i	unfolding dimension
fold	$\mu_1(x)=x^3$	$g_1(x,u)=x^3+ux$	1
cusp	$\mu_2(x)=x^4$	$g_2(x,u,v)=x^4+ux^2+vx$	2
swallowtail	$\mu_3(x)=x^5$	$g_3(x,u,v,w)=x^5+ux^3+vx^2+wx$	3
butterfly	$\mu_4(x)=x^6$	$g_4(x,u,v,w,t)=x^6+ux^4+vx^3+wx^2+tx$	4
hyperbolic umbilic (wave crest)	$\mu_5(x,y)=x^3+y^3$	$g_5(x,y,u,v,w)=x^3+y^3+uxy+vx+wy$	3
elliptic umbilic (hair)	$\mu_6(x,y)=x^3-xy^2$	$g_6(x,y,u,v,w)=x^3-xy^2+u(x^2+y^2)+vx+wy$	3
parabolic umbilic (mushroom)	$\mu_7(x,y)=x^2y+y^4$	$g_7(x,y,u,v,w,t)=x^2y+y^4+ux^2+vy^2+wx+ty$	4
false cusp	$\mu_8(x)=-x^4$	$g_8(x,u,v)=-x^4+ux^2+vx$	2
false butterfly	$\mu_9(x)=-x^6$	$g_9(x,u,v,w,t)=-x^6+ux^4+vx^3+wx^2+tx$	4
false parabolic umbilic (false mushroom)	$\mu_{10}(x)=x^2y-y^4$	$g_{10}(x,u,v,w,t)=x^2y-y^4+ux^2+vy^2+wx+ty$	4

The pairs g_2 and g_8; also g_4 and g_9; and g_7 and g_{10} are equivalent. Except for these three pairs, no two of the unfoldings in the list are similar, nor are any of them similar to the trivial unfolding $0 \in m(0)$. Moreover, no two of these unfoldings are orientedly equivalent.

Remark: If an unfolding f reduces orientedly to one of the unfoldings g_i in the list of theorem 5.7, then in fact g_i is uniquely determined. This does not follow trivially from the theorem as stated above, because although no two of the g_i are orientedly equivalent, they could conceivably become so after quadratic forms in new variables are added. That this cannot actually happen follows from the more general result that an irreducible right-reduction of a germ in m(n) is uniquely determined up to right-equivalence. The proof of this result is fairly long and complicated, so we do not give it here. Thom outlines a proof in his book [15,§5.2D]; Thom's proof depends on a theorem of Tougeron's [17, Proposition 2, p.209]. Incidentally, for theorem 5.6 the corresponding uniqueness statement is trivial, because no two unfoldings in the list of theorem 5.6 are similar, as we shall prove below.

Note: For this theorem too the hypothesis that f have local minima near 0 plays no role in the classification but merely ensures that the reductions have index 0.

It is clear that every unfolding in the expanded list is stable (since they are all equivalent to unfoldings in the list of Theorem 5.6), and they all have local minima near 0, so they can all actually occur in the classification.

Of course here too, if an unfolding f reduces to an unfolding in the list of Theorem 5.7, the index of the reduction is uniquely determined.

Remark on the interpretation of these two theorems for Thom's catastrophe theory:

Thom always states his list in the form of Theorem 5.6 and never in the form of Theorem 5.7, although the information about minima which plays a major role in Thom's theory of natural phenomena is preserved only by oriented reductions. The reason is that Theorem 5.6 and Theorem 5.7 are trivial corollaries of each other, so Theorem 5.6 does already contain all the information one needs. The proper way to interpret Thom's list in its usual form, i.e. in the form of Theorem 5.6, is as an abbreviation for theorem 5.7.

Theorem 5.6 is also perfectly adequate, in fact more convenient than 5.7, for determining the structure of the loci of minima which can arise in the applications of Thom's theory (recall that the locus of minima of an unfolding f consists of the points $(x,u) \in \mathbb{R}^{n+r}$ where $f|\mathbb{R}^n \times \{u\}$ has a local minimum). For it is no more work to find both the locus of minima and the locus of maxima of an unfolding g_i in the abbreviated list than it is to find the minima alone. All investigations done thus far of the structure of the unfoldings in Thom's list in fact include information about the maxima. And the locus of minima of $-g_i$ is of course just the locus of maxima of g_i.

The proof of theorems 5.6 and 5.7 will occupy the rest of this chapter, and will be divided into several lemmas.

We begin with some basic lemmas about reductions.

Lemma 5.8. (a) Let $\eta \in m(p)^2$ and let $\mu \in m(n)^2$. If η and μ are similar, then they have the same right-codimension and the same right-left codimension.

(b) Let $f \in \mathcal{E}(p+b)$ unfold $\eta \in m(p)$. Let $g \in \mathcal{E}(n+r)$ unfold $\mu \in m(n)$. If f and g are similar then f is stable if and only if g is stable.

Proof: Clearly it is enough to prove part (a) for the special case when $p = n+q$ and η reduces on both sides to μ, and it is enough to prove part (b) for the special case when $p = n+q$ and $b = r+s$ and f reduces to g.

Step 1: First we establish a fact which will be used in the proofs of both parts. Let $\mu \in m(n)$ and suppose $\nu \in m(n+q)$ has the form

$$(*) \qquad \nu(x,y) = \mu(x) \pm y_1^2 \pm \dots \pm y_q^2 \qquad (x \in R^n,\ y \in R^q)$$

for some choice of the $\pm$ signs. Note that $\mathcal{E}(n+q) = \mathcal{E}(n)+m(q)\mathcal{E}(n+q)$ (by Lemma 1.6). From this and from (*) it follows that

$$(**) \quad \langle \partial\nu/\partial x_1, \dots, \partial\nu/\partial x_n, \partial\nu/\partial y_1, \dots, \partial\nu/\partial y_q \rangle_{\mathcal{E}(n+q)}$$

$$= \langle \partial\mu/\partial x_1, \dots, \partial\mu/\partial x_n \rangle_{\mathcal{E}(n+q)} + \langle y_1, \dots, y_q \rangle_{\mathcal{E}(n+q)}$$

$$= \langle \partial\mu/\partial x_1, \dots, \partial\mu/\partial x_n \rangle_{\mathcal{E}(n)} + \langle \partial\mu/\partial x_1, \dots, \partial\mu/\partial x_n \rangle_{m(q)\mathcal{E}(n+q)} + m(q)\mathcal{E}(n+q)$$

$$= \langle \partial\mu/\partial x_1, \dots, \partial\mu/\partial x_n \rangle_{\mathcal{E}(n)} + m(q)\mathcal{E}(n+q) .$$

Moreover, suppose ω is an element of $\mathcal{E}(1)$. We may consider $\mu^*(\omega)$ as an element of $\mathcal{E}(n+q)$. Then by (*) it follows that for $x \in R^n$, we have:

$$(\nu^*(\omega)-\mu^*(\omega))(x,0) = \omega\nu(x,0)-\omega\mu(x) = \omega\mu(x)-\omega\mu(x) = 0 .$$

So $\nu^*(\omega)-\mu^*(\omega) \in m(q)\mathcal{E}(n+q)$ (by lemma 1.4).

Hence

$$\nu^*\mathcal{E}(1)+m(q)\mathcal{E}(n+q) = \mu^*\mathcal{E}(1)+m(q)\mathcal{E}(n+q)$$

and from this and (**) it follows that

$$(***) \quad \langle \partial\nu/\partial x_1,\dots,\partial\nu/\partial y_q\rangle_{\mathcal{E}(n+q)}+\nu^*\mathcal{E}(1) = \langle \partial\mu/\partial x_1,\dots,\partial\mu/\partial x_n\rangle_{\mathcal{E}(n)} +m(q)\mathcal{E}(n+q)+\nu^*\mathcal{E}(1)$$

$$= \langle \partial\mu/\partial x_1,\dots,\partial\mu/\partial x_n\rangle_{\mathcal{E}(n)}+\mu^*\mathcal{E}(1)+m(q)\mathcal{E}(n+q).$$

Proof of part (a): Suppose $\eta \in m(p)^2$ and $\mu \in m(n)^2$. Suppose $p = n+q$ and μ is a right-left reduction of η. Then there is a $\nu \in m(p)$ of the form (*) such that $\eta \sim_{rl} \nu$.

Now from (**) it is clear that $\tau(\nu) = \tau(\mu)$, and from (***) it is clear that $\sigma(\nu) = \sigma(\mu)$ (recall the definition of τ and σ from Definition 2.14). Hence r-codim(ν) = r-codim(μ) and rl-codim(ν) = rl-codim(μ). But $\nu \sim_{rl} \eta$, and the right and right-left codimension depend only on the rl-equivalence class. So μ and η have the same right codimension and the same right-left codimension.

Proof of part (b): Suppose $f \in \mathcal{E}(n+q+r+s)$ unfolds $\eta \in m(n+q)$ and suppose $g \in \mathcal{E}(n+r)$ unfolds $\mu \in m(n)$. Furthermore, suppose f reduces to g. Then there is an unfolding $h \in \mathcal{E}(n+q+r)$ of the form $h(x,y,u) = g(x,u) \pm y_1^2 \pm \dots \pm y_q^2$ (for some choice of the $\pm$ signs and for $x \in \mathbb{R}^n$, $y \in \mathbb{R}^q$ and $u \in \mathbb{R}^r$) such that f is equivalent to a constant unfolding of h.

Now if $1 \leq i \leq r$ and if $x \in \mathbb{R}^n$ and $y \in \mathbb{R}^q$, then clearly

$$\alpha_i(h)(x,y) = \frac{\partial h}{\partial u_i}(x,y,0) = \frac{\partial g}{\partial u_i}(x,0) = \alpha_i(g)(x).$$

Hence if we consider $\alpha_i(g)$ as an element of $\mathcal{E}(n+q)$, then $\alpha_i(g) = \alpha_i(h)$ $(i=1,\dots,r)$. Hence $W_g = W_h$ as subspaces of $\mathcal{E}(n+q)$.

Moreover, if we set $\nu = h|\mathbb{R}^{n+q}$, then ν has the form (*). So by (***), and since $W_g = W_h$, it follows that

$$\langle \partial\nu/\partial x_1, \dots, \partial\nu/\partial y_q \rangle_{\mathcal{E}(n+q)} + \nu^* \mathcal{E}(1) + W_h =$$

$$\langle \partial\mu/\partial x_1, \dots, \partial\mu/\partial x_n \rangle_{\mathcal{E}(n)} + \mu^* \mathcal{E}(1) + W_g + m(q)\mathcal{E}(n+q) .$$

Now since $\mathcal{E}(n+q) = \mathcal{E}(n) \oplus m(q)\mathcal{E}(n+q)$ (direct sum as $\mathbb{R}$-vector spaces), it is clear from Theorem 3.22 that h is right-left universal if and only if g is. Since f is equivalent to a constant unfolding of h, f is right-left universal exactly when h is, and hence exactly when g is. Because stability is equivalent to right-left universality we are done.

<u>Lemma 5.9.</u> Let $f \in \mathcal{E}(n+r)$ be a stable unfolding of $\eta \in m(n)$. Suppose η reduces on both sides (on the right) to $\mu \in m(p)$, and suppose $g \in \mathcal{E}(p+s)$ is a stable unfolding of μ of minimal unfolding dimension. Then f reduces (orientedly) to g.

<u>Proof:</u> Let $b = n-p$. Then there is a non-degenerate quadratic form $Q \in m^2(b)$ such that if we define $\nu \in m(n)$ by $\nu(x,y) = \mu(x)+Q(y)$ for $x \in \mathbb{R}^p$ and $y \in \mathbb{R}^b$, then ν and η are

right-left (right) equivalent.

Now let $a = r-s$. Since f is stable and g is stable of minimal unfolding dimension, we have (applying 5.8(a)) that $r \geq \text{rl-codim}(\eta) = \text{rl-codim}(\mu) = s$. Hence $a \geq 0$. Define an unfolding $h \in \mathcal{E}(n+r)$ by setting $h(x,y,u,v) = g(x,u)+Q(y)$ for $x \in \mathbb{R}^p$, $y \in \mathbb{R}^b$, $u \in \mathbb{R}^s$, $v \in \mathbb{R}^a$. Then h unfolds ν and since ν and η are right-left (right) equivalent, it follows that there is an unfolding $h' \in \mathcal{E}(n+r)$ which unfolds η and which is (orientedly) equivalent to h. Now h reduces to g, so by Lemma 5.8(b) h is stable; hence so is h'. But f and h' are stable unfoldings of η of the same dimension, so by Theorem 3.20 they are right-left isomorphic. It follows that f is (orientedly) equivalent to h, so f reduces (orientedly) to g.

<u>Definition 5.10.</u> Let $\eta \in m(n)^2$. We define the <u>corank of η</u> to be the corank of the Hessian matrix H of η at 0. (the Hessian matrix H is an $n \times n$ matrix whose i,j-th element is $\frac{\partial^2 \eta}{\partial x_i \partial x_j}(0)$, for $1 \leq i,j \leq n$.)

<u>Remark:</u> The Hessian matrix H is just the matrix , with respect to the standard coordinates on $\mathbb{R}^n$, of the Hessian form of η, which is a symmetric bilinear form on the tangent space of $\mathbb{R}^n$ at 0. Since the Hessian form can be defined invariantly, without choosing coordinates, and since the corank of H is just the corank of the Hessian form (considered as a linear map from the tangent space to its dual), it follows that the corank of H is invariant under change of coordinates, and hence that the

corank of η is invariant under right-equivalence. Moreover if ω is an element of $L(1)$, then the Hessian matrix of $\omega\eta$ is just H multiplied by a non-zero real number (namely $\partial\omega/\partial t\ (0)$), so it has the same corank as H. Hence the corank of η depends only on the right-left equivalence class of η.

<u>Lemma 5.11.</u> Let $\eta \in m(p)^2$ and let $\mu \in m(n)^2$. Suppose η is similar to μ. Then

$$\operatorname{corank}(\eta) = \operatorname{corank}(\mu)$$

<u>Proof:</u> Clearly it is enough to prove the lemma in the case when $p = n+q$ and η reduces on both sides to μ. Suppose this is the case. Then there is a germ $\nu \in m(p)$, of the form $\nu(x,y) = \mu(x) \pm y_1^2 \pm \ldots \pm y_q^2$ $(x \in \mathbb{R}^n,\ y \in \mathbb{R}^q)$ such that $\eta \sim_{rl} \nu$.

Now let H be the Hessian matrix of μ, and let H' be the Hessian matrix of ν. Then H' clearly has the form

$$H' = \begin{array}{c} n\{ \\ q\{ \end{array} \overbrace{\left(\begin{array}{c|ccc} H & & 0 & \\ \hline & \pm 2 & & 0 \\ 0 & & \ddots & \\ & 0 & & \pm 2 \end{array}\right)}^{\;n \qquad\qquad q}$$

so the corank of H' is the same as the corank of H. Hence $\operatorname{corank}(\mu) = \operatorname{corank}(\nu)$. But the corank of a germ depends only on the rl-equivalence class, so $\operatorname{corank}(\nu) = \operatorname{corank}(\eta)$, and we are done.

<u>Lemma 5.12.</u> (Splitting Lemma) Let $\eta \in m(n)^2$, and let p be the corank of η. Then η reduces on the right to a germ $\mu \in m(p)^3$

<u>Remark:</u> In fact μ is uniquely determined up to right equivalence, but we do not prove this here. Thom sketches a proof in [15,§ 5.2D]. See also the remark following the statement of Theorem 5.7.

<u>Proof of Lemma 5.12:</u> Let $q = n-p$. We consider $\mathbb{R}^n$ as $\mathbb{R}^p \times \mathbb{R}^q$ and take coordinates $x_1, \dots, x_p$, $y_1, \dots, y_q$ on $\mathbb{R}^n$.

We first show that η is right-equivalent to a germ in $m(n)^2$ whose restriction to $\mathbb{R}^q$ is a non-degenerate quadratic form in $y_1, \dots, y_q$.

As we have remarked, the Hessian matrix of η is just the matrix, with respect to the standard coordinates on $\mathbb{R}^n$, of the Hessian form of η, which is a symmetric bilinear form on the tangent space of $\mathbb{R}^n$ at 0. Now it follows from elementary linear algebra that with respect to suitable coordinates on $\mathbb{R}^n$ the matrix of this symmetric bilinear form will be diagonal and of the form

$$\begin{pmatrix} 0 & & & & & & & & \\ & \ddots & & & & & 0 & & \\ & & 0 & & & & & & \\ & & & -2 & & & & & \\ & & & & \ddots & & & & \\ & & & & & -2 & & & \\ & & 0 & & & & +2 & & \\ & & & & & & & \ddots & \\ & & & & & & & & +2 \end{pmatrix}$$

In other words there is a germ $\varphi_1 \in L(n)$ such that the Hessian matrix of $\eta\varphi_1$ has this form. Since the corank of this matrix is the corank of η, i.e. p, there must be p zeroes along the diagonal. Let λ be the number of -2's, and define $Q \in m(q)^2$ by

$$Q(y) = -y_1^2-\ldots-y_\lambda^2+y_{\lambda+1}^2+\ldots+y_q^2 \quad (y \in R^q).$$

Now define a germ $\nu \in m(q)$ by

$$\nu(y) = \eta\varphi_1(0,y) \quad \text{for } y \in R^q \quad (\text{here } 0 \text{ is the origin of } R^p).$$

Clearly ν and Q have the same Hessian matrix at 0. Moreover $Q \in m(q)^2$; also, since $\eta \in m(n)^2$ it follows that $\eta\varphi_1 \in m(n)^2$ and hence that $\nu \in m(q)^2$. Therefore $\pi_2(\nu) = \pi_2(Q)$. But Q is right 2-determined, so $\nu \sim_r Q$. Pick a germ $\psi \in L(q)$ such that $\nu\psi = Q$. Define $\varphi_2 \in L(n)$ by $\varphi_2(x,y) = \varphi_1(x,\psi(y))$. ($x \in R^p$, $y \in R^q$). Then for $y \in R^q$ we have $\eta\varphi_2(0,y) = Q(y)$. Hence we may consider $\eta\varphi_2$ as a p-dimensional unfolding of Q. But it is easy to see, using theorem 3.22, that Q is a right-universal unfolding of itself. Hence $\eta\varphi_2$ is right-isomorphic to the p-dimensional constant unfolding of Q. Pick a right-isomorphism (Φ,α), with $\Phi \in L(n)$ and $\alpha \in m(p)$, such that

$$Q(y) = \eta\varphi_2\Phi(x,y) + \alpha(x) \quad \text{for } x \in R^p,\ y \in R^q .$$

Set $\mu := -\alpha$; from the equation above we see that η reduces on the right to $\mu \in m(p)$. Moreover $\mu \in m(p)^2$, since $Q \in m(q)^2$ and $\eta\varphi_2\Phi \in m(n)^2$. And by Lemma 5.11, corank (μ) = corank (η) = p, so the Hessian matrix of μ must be 0, for otherwise the corank would be less than the dimension of the Euclidean space on which μ is defined, i.e. the corank would be less than p. It follows that $\mu \in m(p)^3$, and we are done.

As noted in the introduction, Mather gives an entirely different proof in [10]. For his proof, Mather requires that η be finitely determined. Gromoll and Meyer prove a more general (Hilbert-space) version of this lemma in [2].

<u>Corollary 5.13.</u> (a) Let $\nu \in m(p)^2$. Then ν is irreducible if and only if $\nu \in m(p)^3$.
(b) Let $g \in \mathcal{E}(p+r)$ be a stable unfolding of $\nu \in m(p)^2$. Then g is irreducible if and only if $\nu \in m(p)^3$ and the right-left codimension of ν is r.

<u>Proof:</u> (a) By lemmas 5.12 and 5.11 it is clear that ν is irreducible if and only if corank $(\nu) = p$ (note that corank (ν) is always $\leq p$). But corank $(\nu) = p$ if and only if the Hessian of ν vanishes, i.e. if and only if $\nu \in m(p)^3$.

(b) "if" Suppose $f \in \mathcal{E}(q+s)$ unfolds $\mu \in m(q)$ and g reduces to f. Then ν reduces on both sides to μ. Since $\nu \in m(p)^3$, clearly ν is irreducible, so q=p. Since g reduces to f, f is also stable, so $s \geq$ rl-codim μ = rl-codim ν = r. Hence s=r, so g is irreducible.
"only if". Suppose g is irreducible. Now let $\mu \in m(q)$ be any right-left reduction of ν. Since g is stable it follows that ν and hence also μ are finitely determined. So we can find an unfolding $f \in \mathcal{E}(q+s)$, for some s, which unfolds μ and which is stable of minimal unfolding dimension. By Lemma 5.9 g reduces to f. Since g is irreducible, we have r = s = rl-codim μ = rl-codim ν, and we have q = p, which means ν is irreducible, so by part (a) it follows that $\nu \in m(p)^3$. This completes the proof.

Corollary 5.14 (a) Suppose $\eta \in m(n)^2$ reduces on both sides to an irreducible germ $\mu \in m(p)^2$. Then the index of the reduction is uniquely determined, and in fact depends only on η, not on μ.

(b) Suppose $f \in \mathcal{E}(n+r)$ unfolds $\eta \in m(n)$. Suppose $g \in \mathcal{E}(p+s)$ is a stable and irreducible unfolding of $\mu \in m(p)^2$, and suppose f reduces to g. Then the index of the reduction is uniquely determined, and in fact depends only on f, not on g.

Proof: (a) Let $q = n-p$. Suppose η reduces on both sides to μ with index λ. Then for some choice of the $\pm$ sign, if we set $\nu(x,y) = \pm\mu(x)-y_1^2- \dots -y_\lambda^2 + y_{\lambda+1}^2 + \dots + y_q^2 (x \in R^p,\ y \in R^q)$, then there is an $\omega \in L(1)$ with $\frac{d\omega}{dt}(0) > 0$ such that $\omega\eta \sim_r \nu$. Now since $\frac{d\omega}{dt}(0) > 0$, the index of the Hessian of η equals the index of the Hessian of $\omega\eta$, and since $\omega\eta \sim_r \nu$, this equals the index of the Hessian of ν, which is clearly λ plus the index of the Hessian of $\pm\mu$. But since μ is irreducible, it follows that μ, and hence also $-\mu$, are in $m(p)^3$, so the Hessian of $\pm\mu$ is 0. Hence λ is the index of the Hessian of η and so is uniquely determined and independent of μ.

(b) Since g is stable, and since g is irreducible, μ is irreducible by Corollary 5.13. If f reduces to g with index λ, then η reduces to μ with index λ, so by part (a), λ is uniquely determined and depends only on η, hence not on g.

The next step in proving theorems 5.6 and 5.7 is to classify the germs of right-left codimension ≤ 4.

Theorem 5.15. (Classification theorem) Let $\eta \in m(n)^2$ and suppose the right-left codimension of η is ≤ 4. Then η reduces on the right to one of the following germs ν_i:

ν_i	rl-codim ν_i	corank ν_i
ν_1 is the unique germ in $m(0)$	0	0
$\nu_2(x) = x^3$	1	1
$\nu_3(x) = x^4$	2	1
$\nu_4(x) = -x^4$	2	1
$\nu_5(x) = x^5$	3	1
$\nu_6(x) = x^6$	4	1
$\nu_7(x) = -x^6$	4	1
$\nu_8(x,y) = x^3+y^3$	3	2
$\nu_9(x,y) = x^3-xy^2$	3	2
$\nu_{10}(x,y) = x^2y+y^4$	4	2
$\nu_{11}(x,y) = x^2y-y^4$	4	2

All of these germs are right-left irreducible. The germs ν_3 and ν_4 are right-left equivalent, as are the germs ν_6 and ν_7 and the germs ν_{10} and ν_{11}; but no other pairs are similar.

The proof we give (except for the proof of the uniqueness statement) is essentially taken from Mather [10, Chapter II]. Mather classifies the germs of right-codimension ≤ 5, but the same proof works here because for the germs of rl-codimension ≤ 4 the right and right-left codimensions are the same.

We have only stated here what we need to prove the "Thom theorem", and one can in fact prove much more. For example, Mather shows in addition that the set of 7-jets of germs in $m(n)$

of right-codimension > 5 is an algebraic subset of $J^7_o(n,1)$ of codimension n+6. And Siersma [12] has extended Mather's classification to germs of right-codimension ≤ 8. The same methods should work to classify also the germs of right-left codimension ≤ 8, since in most cases the right and right-left codimensions are the same (although Siersma's list contains a class of germs of right-left codimension 7 and right codimension 8; this is in fact the earliest case in which the two codimensions can differ).

For the germs $\nu_o,\dots,\nu_{11}$ given above, one can easily verify by computation that the right and right-left codimensions are the same.

<u>Proof of Theorem 5.15.</u> We recall that rl-codim $(\eta) = \sigma(\eta)$. To prove the theorem, we examine all the cases which can occur.

Remark: If η reduces on the right to $\mu \in m(q)$ and if $\mu' \in m(q)$ and $\mu \sim_r \mu'$, then clearly μ' is also a right-reduction of η.

<u>Case A.</u> Corank $(\eta) = 0$. Then by lemma 5.12, η reduces on the right to ν_1.

<u>Case B.</u> Corank $(\eta) = 1$. Then η reduces on the right to a germ $\mu \in m(1)^3$. Now by Lemma 5.8(a), we also have $\sigma(\mu) \leq 4$, and in particular μ is finitely determined. It follows that $\mu \notin m(1)^\infty$, and hence that for some integer $k \geq 3$ and for some $\omega \in m(1)$ and some real number $a > 0$, we have $\mu(x) = \pm(ax^k+x^k\omega(x))$. If we define a germ φ by $\varphi(x) = x\sqrt[k]{a+\omega(x)}$ then $\varphi \in L(1)$ and $\mu(x) = \pm(\varphi(x))^k$. So $\mu \sim_r \nu$, where $\nu(x) = \pm x^k$. Now one easily

computes that rl-codim $\nu = k-2$ and hence k must be 3,4,5, or 6 (since we must have rl-codim $(\nu) \leq 4$). Moreover, if k is odd, then $\nu \sim_r -\nu$, so we may assume $\nu(x) = +x^k$ (if k odd). Hence μ is right-equivalent to one of the germs $\nu_2, \nu_3, \dots, \nu_7$ and so η reduces on the right to one of these germs.

Case C. Corank $(\eta) = 2$. Then η reduces on the right to a germ $\mu \in m(2)^3$. Set $P = \pi_3(\mu)$; then P is a homogeneous cubic polynomial with real coefficients in two variables x and y. Now P/y^3 is a polynomial (with real coefficients) in x/y, and P/y^3 has degree at most 3. Since P/y^3 splits into linear factors over $\mathbb{C}$, it follows that P splits into linear factors over $\mathbb{C}$, i.e. we can write

$$P(x,y) = (a_1x+b_1y)(a_2x+b_2y)(a_3x+b_3y)$$

where the a_i and b_i are complex numbers. Four cases can occur:

Case C1. The three vectors $(a_i,b_i) \in \mathbb{C}^2$ are pairwise linearly independent over $\mathbb{C}$.

Case C2. Two of the vectors (a_i,b_i) are linearly independent over $\mathbb{C}$, but the third is a non-zero multiple of one of the others. In this case we may write $P(x,y) = (a_1'x+b_1'y)(a_2'x+b_2'y)^2$, where (a_1',b_1') and (a_2',b_2') are linearly independent. Note that this implies that the polynomial P/y^3 is either of degree 1 or has a double root; in either case all the roots of P/y^3 are real, so we may take the a_i' and b_i' to be real.

Case C3. No two of the vectors (a_i,b_i) are linearly independent, but neither is any of them zero. In this case we may write

$$P(x,y) = (ax+by)^3 \neq 0 .$$

Here too we may take a and b to be real.

<u>Case C4.</u> $P = 0$.

Proof for Case C1:

Case C1a: All the roots of the polynomial P/y^3 are real. In this case we may take a_i and b_i $(i=1,2,3)$ all to be real. If we take a_1x+b_1y and a_2x+b_2y as new coordinates on $\mathbb{R}^2$, we see $P(x,y) \sim_r xy(ax+by)$ for some real numbers a and b. Now since the factors are pairwise linearly independent linear forms, a and b are both nonzero. So clearly $xy(ax+by) \sim_r \frac{1}{ab} xy(x+y) \sim_r xy(x+y)$. Take $x+y$ and $x-y$ as new coordinates. Then one sees $xy(x+y) \sim_r x(x^2-y^2) = x^3-xy^2$. Since $P = \pi_3(\mu)$ it follows that there is a $\mu' \in m(2)$ with $\mu \sim_r \mu'$ and $\pi_3(\mu') = x^3-xy^2$. But using theorem 2.6 one easily sees that x^3-xy^2 is right 3-determined. Hence $\mu \sim_r x^3-xy^2$, and so η reduces on the right to $x^3-xy^2 = \nu_9$.

Case C1b: Two of the roots of P/y^3 are not real. They must then be complex conjugates of each other, and the third root (if it exists, i.e. if P/y^3 has degree 3) must be real. So we may write $P(x,y) = (a_1x+b_1y)(a_2x+b_2y)(\overline{a}_2x+\overline{b}_2y)$. Now $Q(x,y) := (a_2x+b_2y)(\overline{a}_2x+\overline{b}_2y)$ is a positive definite quadratic form on $\mathbb{R}^2$ (it is definite because the two factors are linearly independent). In suitable coordinates Q becomes x^2+y^2, so we see that $P \sim_r (ax+by)(x^2+y^2)$. Now by an orthogonal change of coordinates we can arrange that $ax+by$ becomes cx $(c \neq 0)$ and of course the form x^2+y^2 does not change. So $P \sim_r cx(x^2+y^2) \sim_r x(x^2+y^2) = x^3+xy^2$. Now $x^3+xy^2 \sim_r x^3+y^3$, for if one replaces x by $x+y$ and y by $x-y$ in x^3+y^3, one sees that $x^3+y^3 \sim_r 2x^3+6xy^2 \sim_r x^3+xy^2$. Hence $P \sim_r x^3+y^3$. But one easily sees that x^3+y^3 is right 3-determined. So $\mu \sim_r x^3+y^3$ and η reduces on the right to $x^3+y^3 = \nu_8$.

Case C2: $P(x,y) = (a_1x+b_1y)(a_2x+b_2y)^2 \sim_r x^2y$. Now x^2y is not finitely determined, because $\partial x^2y/\partial x = 2xy$, and $\partial x^2y/\partial y = x^2$, and no power of y lies in the ideal $\langle 2xy, x^2\rangle$. Hence we must look at higher-order jets of μ. It cannot be the case that $\pi_k(\mu) \sim_r x^2y$ for arbitrarily high k, for μ is finitely determined, so $\pi_k(\mu)$ must be right k-determining for sufficiently large k. So let k be maximal, such that $\pi_k(\mu) \sim_r x^2y$. (Note that then $k \geqslant 3$, since $\pi_3(\mu) \sim_r x^2y$). We shall show that $\mu \sim_r x^2y \pm y^{k+1}$. Choose a germ $\mu' \in m(2)$ such that $\mu \sim_r \mu'$ and such that $\pi_k(\mu') = x^2y$. Then $\pi_{k+1}(\mu') = x^2y+h(x,y)$, where h is a homogeneous polynomial of degree $k+1$ in x and y. Consider a map $\Phi: R^2 \longrightarrow R^2$ of the form $\Phi(x,y) = (x+\varphi(x,y), y+\psi(x,y))$ where φ and ψ are homogeneous polynomials in x and y of degree $k-1$. Since $k \geqslant 3$, Φ is a local diffeomorphism of R^2 near the origin. And $\pi_{k+1}(\mu'\Phi) = x^2y+x^2\psi+2xy\varphi+h(x,y)$ (all other terms which occur on the right after substituting $x+\varphi$ for x and $y+\psi$ for y are of degree $\geqslant k+2$, since φ and ψ have degree $k-1$ and $k \geqslant 3$.) Now by choosing φ and ψ appropriately, we can clearly kill all terms of $h(x,y)$ which are divisible by xy or by x^2. Hence $\pi_{k+1}(\mu') \sim_r x^2y+ay^{k+1}$ for some real number a. By the choice of k, $a \neq 0$. One easily sees, using Theorem 2.6, that x^2y+ay^{k+1} is right $k+1$-determined, so $\mu \sim_r x^2y+ay^{k+1}$. But clearly $x^2y+ay^{k+1} \sim_r x^2y \pm y^{k+1}$, where the sign depends on whether a is positive or negative.

Now if $k \geq 4$, one easily computes that $\sigma_5(x^2y \pm y^{k+1}) \geq 5$. Since $\sigma(\mu) \leq 4$, we must have $k = 3$, and it follows that η reduces on the right to ν_{10} or to ν_{11}.

Case C3: $P(x,y) = (ax+by)^3 \sim_r x^3$. Pick $\mu' \in m(2)$ such that $\mu' \sim_r \mu$ and such that $\pi_3(\mu') = x^3$. Then $\pi_4(\mu') = x^3+h(x,y)$, where h is a homogeneous polynomial of degree 4 in x and y. Clearly $\pi_3(\langle \partial\mu'/\partial x, \partial\mu'/\partial y\rangle + \mu'^*\mathcal{E}(1))$ is generated by the 3-jets 1, $3x^2+\partial h/\partial x$, $\partial h/\partial y$, x^3 and x^2y, i.e. by five elements. But $\dim_{\mathbb{R}} J^3(2,1) = 10$, so $\sigma_4(\mu') \geq 10-5 = 5$. Hence this case cannot occur.

Case C4: $P = 0$, i.e. $\mu \in m(2)^4$. Then one easily computes $\sigma_3(\mu) = 5$, so this case cannot occur.

Case D. Corank $(\eta) \geq 3$. Let p = corank (η). Then η reduces on the right to a germ $\mu \in m(p)^3$. Let $h = \pi_3(\mu)$. Then $\pi_2(\langle \partial\mu/\partial x_1, \ldots, \partial\mu/\partial x_p\rangle + \mu^*\mathcal{E}(1))$ is generated by the p+1 two-jets $\partial h/\partial x_1, \ldots, \partial h/\partial x_p$ and 1. But $\dim_{\mathbb{R}} J^2(p,1) = \binom{p+2}{2} = \frac{1}{2}(p+1)(p+2)$, as one easily sees. So $\sigma_3(\mu) \geq \frac{1}{2}(p+1)(p+2)-(p+1) = \frac{1}{2}(p^2+p) \geq 6$, since $p \geq 3$. Hence this case cannot occur.

Clearly, we have now considered all possible cases. It is an easy computation to check the rl-codimensions given in the table for the germs $\nu_1, \ldots, \nu_{11}$. Moreover, since $\pi_2(\nu_i) = 0$ for each ν_i, it is clear that the coranks given in the table are correct, and it also follows that all the ν_i are irreducible (by Corollary 5.13). The germ ν_{10} is obviously right-equivalent to $-x^2y+y^4 = -\nu_{11}$, so ν_{10} and ν_{11} are similar (in fact right-left equivalent). And obviously $\nu_3 \sim_{rl} \nu_4$ and $\nu_6 \sim_{rl} \nu_7$. It follows from Lemma 5.8(a) and from Lemma 5.11 that no other pair, except possibly ν_8 and ν_9, can be similar, for except for this pair and the ones mentioned above, no two of the germs in the list have both the same rl-codimension and the same corank. So it only remains to show that ν_8 and ν_9 are not similar.

If $\eta \in m(n)^2$ we shall say η has property A if there exists a germ X at 0 of a smooth vector field on $\mathbb{R}^n$, such that $X(0) \neq 0$ and such that the germ $X(\eta) \in m(n)$ has either a local maximum or a local minimum at 0. Of course, this is the same as saying $X(\eta)$ is either non-positive near 0 or non-negative near 0. Such a germ X will be called an A-vector field for η.

We shall see that ν_8 has property A and ν_9 does not. But first, let us show that property A is a similarity invariant for germs in $m(n)^2$.

<u>Remark:</u> One easily sees that singularity (i.e. the property of being in $m(n)^2$) is a similarity invariant. That is, if $\eta \in m(n)^2$ and η is similar to $\mu \in m(q)$, then $\mu \in m(q)^2$.

<u>Step 1.</u> Let $\mu \in m(n)^2$, and let $\nu \in m(n+q)^2$ have the form $\nu(x,y) = \mu(x) \pm y_1^2 \pm \dots \pm y_q^2$ $(x \in \mathbb{R}^n,\ y \in \mathbb{R}^q)$. Then ν has property A if and only if μ does. First suppose X is an A-vector field for μ. We have a canonical projection $\pi\colon \mathbb{R}^{n+q} \longrightarrow \mathbb{R}^n$ and a canonical imbedding ι of the tangent bundle of $\mathbb{R}^n$ in the tangent bundle of $\mathbb{R}^{n+q}$. Define a germ X' of a vector field on $\mathbb{R}^{n+q}$ by setting $X' = \iota X \pi$. Then $X'(0) = \iota X(0)$ and $X'(\nu) = X(\mu)\pi$, so clearly X' is

an A-vector field for ν. Now suppose Y is an A-vector field for ν and write $Y(x,y) = \sum_{i=1}^{n} \beta_i(x,y)\frac{\partial}{\partial x_i} + \sum_{j=1}^{q} \gamma_j(x,y)\frac{\partial}{\partial y_j}$,

where the β_i and γ_j are in $\mathcal{E}(n+q)$. Define a germ Y' of a vector field on $\mathbb{R}^n$ by $Y'(x) = \sum_{i=1}^{n} \beta_i(x,0)\frac{\partial}{\partial x_i}$.

Now if $Y'(0) = 0$, then since $Y(0,0) \neq 0$ there must be a j, $1 \leq j \leq q$, such that $\gamma_j(0,0) \neq 0$. But then clearly $Y(\nu)$ has linear terms in the y_j's and hence $Y(\nu) \notin m(n+q)^2$. This means that $Y(\nu) \sim_r y_1$, for example, and hence cannot have a minimum or maximum at 0. So $Y'(0) \neq 0$. Since $Y'(\mu) = Y(\nu)|\mathbb{R}^n$, clearly $Y'(\mu)$ has a minimum or maximum at 0, so Y' is an A-vector field for μ. This completes step 1.

<u>Step 2.</u> Let $\mu \in m(n)^2$ and $\eta \in m(n)^2$ and suppose $\mu \sim_r \eta$. Suppose η has property A. Choose $\varphi \in L(n)$ such that $\eta = \mu\varphi$. Let X be an A-vector field for η. Now φ induces a germ $d\varphi$ of a bundle map, lying over φ, of the tangent bundle of $\mathbb{R}^n$ to itself. Set $X' = d\varphi(X)$. Since $d\varphi$ is an isomorphism on each fibre and since $\varphi(0) = 0$, clearly $X'(0) \neq 0$. And we have $X'(\mu) = X(\mu\varphi) = X(\eta)$, so clearly $X'(\mu)$ also has a local minimum or maximum at 0. Hence μ also has property A.

<u>Step 3.</u> Let $\mu \in m(n)^2$ and $\eta \in m(n)^2$. Suppose η has property A, and suppose there exists $\omega \in L(1)$ such that $\mu = \omega\eta$. Let X be an A-vector field for η. Then $X(\eta)$ is either everywhere non-negative or everywhere non-positive. Now for each x near 0 in $\mathbb{R}^n$, $X(\mu)(x) = X(\omega\eta)(x) = \frac{\partial\omega}{\partial t}(\eta(x))\cdot X(\eta)(x)$. Since $\frac{\partial\omega}{\partial t}(\eta(0)) = \frac{\partial\omega}{\partial t}(0) \neq 0$, clearly also $X(\mu)$ is either everywhere non-negative or everywhere non-positive near $0 \in \mathbb{R}^n$. Hence X is also an A-vector field for μ.

From Steps 1-3 it is clear that property A depends only on the similarity class of a germ.

Now ν_8 obviously has property A, for $\frac{\partial \nu_8}{\partial x}(x,y) = 3x^2$, which has a local minimum at 0. We shall show ν_9 does not have property A. For suppose X is an A-vector field for ν_9. Write

$X = \beta(x,y)\frac{\partial}{\partial x} + \gamma(x,y)\frac{\partial}{\partial y}$, where β and γ are in $\mathcal{E}(2)$.

Then $X(\nu_9(x,y)) = \beta(x,y)(3x^2-y^2)+\gamma(x,y)(2xy)$.

First suppose $\gamma(0,0) \neq 0$. Then we may assume without loss of generality that $\gamma(0,0) > 0$ and hence $\gamma(x,y) > 0$ everywhere near 0, for if this is not the case, we may replace X by -X, which is also an A-vector field for ν_9. Now observe that for all x near 0 but unequal to 0, we have

$X(\nu_9)(x,+\sqrt{3}x) = \gamma(x,\sqrt{3}x)(2\sqrt{3}x^2) > 0$, and

$X(\nu_9)(x,-\sqrt{3}x) = \gamma(x,-\sqrt{3}x)(-2\sqrt{3}x^2) < 0$. Since X is an A-vector field for ν_9, this is a contradiction, so $\gamma(0,0) = 0$. But since $X(0,0) \neq 0$, we must then have $\beta(0,0) \neq 0$. In other words, we have $\beta \notin m(2)$ and $\gamma \in m(2)$. Hence $\pi_2(X(\nu_9)) = \beta(0,0)(3x^2-y^2)$ $\sim_r \pm(x^2-y^2)$. But clearly $\pm(x^2-y^2)$ is right 2-determined, by Theorem 2.6. So $X(\nu_9) \sim_r \pm(x^2-y^2)$. But clearly $\pm(x^2-y^2)$ has neither a maximum nor a minimum at 0, so neither has $X(\nu_9)$. Hence no A-vector field for ν_9 exists, and consequently ν_9 is not similar to ν_8.

This completes the proof of theorem 5.15.

Remark: The proof that ν_8 and ν_9 are not similar can be shortened considerably if one uses the fact, remarked upon after Lemma 5.12, that an irreducible right-reduction of a germ in $m(n)^2$ is uniquely determined up to right-equivalence, and the fact, remarked upon after Definition 5.2, that a germ η reduces on both sides to a germ μ if and only if η reduces on the right to a germ which is rl-equivalent to μ. For then it follows that ν_8 and ν_9 are similar if and only if they are rl-equivalent. And one can easily check that ν_8 and ν_9 are not rl-equivalent; because if they were, then since they are both homogeneous cubic polynomials in x and y, one could find linear diffeomorphisms $\varphi \in L(2)$ and $\omega \in L(1)$ such that $\nu_8 = \omega\nu_9\varphi$; but this is impossible because the cubic form ν_9 factors into a product of three linear forms over $\mathbb{R}$, and ν_8 doesn't; this property is clearly invariant under a linear change of coordinates.

We gave a longer, more direct proof here because we have not proved the uniqueness of irreducible reductions.

Remark: It is not very hard to see that no two of the germs $\nu_1,\ldots,\nu_{11}$ are right-equivalent (the only slight difficulty is in the case of ν_{10} and ν_{11}). We shall prove this fact below, during the proof of theorem 5.7.

We can now complete the proof of the validity of Thom's list.

Proof of theorems 5.6 and 5.7. Let f and η satisfy the hypotheses of theorems 5.6 and 5.7. We shall prove Theorem 5.7; Theorem 5.6 is a trivial corollary.

Now since f is stable, and hence right-left universal, and since the unfolding dimension of f is ≤ 4, it follows that rl-codim $(\eta) \leq 4$. Hence, by theorem 5.15 we know that η reduces on the right to a germ μ which is one of the germs

$\nu_1,\dots,\nu_{11}$ listed in the statement of theorem 5.15. Now ν_1 is the trivial germ $0 \in m(0)$, and if $1 < i \leqslant 11$, then ν_i is one of the germs μ_j given in the list of theorem 5.7.

If $\mu = \nu_1$ then clearly f has a simple singularity at 0. In this case set $g = 0 \in m(0)$. If μ is one of the germs μ_j listed in the statement of theorem 5.7, let $g = g_j$. One easily computes, using theorem 3.22, that in every case g is a stable unfolding of μ and in fact is stable of minimal unfolding dimension, since the unfolding dimension of g is in every case just the right-left codimension of μ as given in the table of Theorem 5.15. So by Lemma 5.9 it follows that f reduces orientedly to g.

Now we must show that f reduces to g with index 0. (if $\mu = \nu_1$ this implies f has a simple minimum at 0). Suppose f reduces to g with index λ. Suppose $\mu \in \mathcal{E}(p)$ and $g \in \mathcal{E}(p+s)$. Let $q = n-p$ and let $b = r-s$. Define $h \in \mathcal{E}(n+r)$ by

$$(*) \quad h(x,y,u,v) = g(x,u)-y_1^2-\dots-y_\lambda^2+y_{\lambda+1}^2+\dots+y_q^2$$

($x \in \mathbb{R}^p$, $y \in \mathbb{R}^q$, $u \in \mathbb{R}^s$, $v \in \mathbb{R}^b$). Then f is orientedly equivalent to h. Since f has local minima near 0 it is clear that h has local minima near 0, for this property is obviously invariant under oriented equivalence.

Choose representatives H of h and G of g such that (*) holds also for these representatives. It follows from Definition 5.5 that there is a point (x',y',u',v') in $\mathbb{R}^{p+q+s+b} = \mathbb{R}^{n+r}$, such that $H|\mathbb{R}^n\times\{(u',v')\}$ has a local minimum at (x',y',u',v'). If $y' \neq 0$ then $H|\mathbb{R}^n\times\{(u',v')\}$ is non-singular at (x',y',u',v') and hence cannot have a minimum there. So y' must be 0. But then clearly $H|\mathbb{R}^n \times \{(u',v')\}$ can have a local minimum at (x',y',u',v') only if there are no terms $-y_i^2$ in (*), i.e. only if $\lambda = 0$. So f reduces to g with index 0.

This completes the proof of theorem 5.7 except for the uniqueness statement. Now $\mu_2 = -\mu_8$; and $\mu_4 = -\mu_9$; and one easily sees that $\mu_7 \sim_r -\mu_{10}$. So since the g_i are minimal stable unfoldings of the μ_i, clearly the pairs g_2 and g_8; also g_4 and g_9; also g_7 and g_{10} are equivalent. No other pair g_i and g_j $(j \neq i)$ can be similar, since then μ_i and μ_j would be similar, but this cannot be by Theorem 5.15. Neither can any of the g_i be similar to the trivial unfolding $0 \in m(0)$. Clearly g_2 and g_8, and g_4 and g_9 are not orientedly equivalent, for μ_2 and μ_4 have minima at the origin and μ_8 and μ_9 have maxima at the origin. It is a bit more difficult to see that g_7 and g_{10} are not orientedly equivalent. If they were there would be a germ $\varphi \in L(2)$ and a germ $\omega \in L(1)$ such that $\frac{\partial \omega}{\partial t}(0) > 0$ and such that $\mu_7 = \omega\mu_{10}\varphi$, i.e. such that $x^2y+y^4 = \omega(\varphi_1^2\varphi_2-\varphi_2^4)(x,y)$ for $(x,y) \in \mathbb{R}^2$. But if one works out the Taylor series of $\omega\mu_{10}\varphi$ up to degree 4 one finds that if the three-jet of $\omega\mu_{10}\varphi$ equals x^2y, then the coefficient of y^4 in the four-jet of $\omega\mu_{10}\varphi$ must be negative, no matter what φ and ω are (assuming of course that $\frac{\partial \omega}{\partial t}(0) > 0$). Hence g_7 and g_{10} are not orientedly equivalent.

That the unfoldings in this list are irreducible follows immediately by Corollary 5.13.

This completes the proof of theorem 5.7. And theorem 5.6 follows trivially from 5.7, so we are done.

APPENDIX

In this appendix we give a brief exposé of René Thom's catastrophe theory and explain how the mathematics in this paper relates to Thom's theory. (See for example [14],[15],[16]).

The theory of catastrophes is intended to give mathematical models for processes occuring in nature. We shall begin with a very rough description of these models and fill in more and more details as we go on.

We suppose two smooth manifolds B and M are given. In Thom's applications of his theory, the manifold B is usually the physical space-time in which the event or process to be described takes place; in other words, B is an open subset of $\mathbb{R}^4$. Zeeman calls B the control space, and he has applications (for example his description of the beating of the heart, or his explanation of flight-attack behaviour in animals) in which B is not a physical space but rather a space of control parameters which govern the event to be described. In this case there is of course no restriction on the dimension of B, but it is also usually small (≤ 4).

The manifold M is called the state space and is parametrised by the various physical variables (e.g. temperature, pressure, etc.) which are relevant to the process under study. This manifold can be of very high dimension.

We consider $B \times M$ via the projection $\pi: B \times M \longrightarrow B$ as a fibre space over B with fibre M.

To fix a definition, we may say that a <u>process</u> is a subset of $B \times M$. If s is a process and $u \in B$, we define s_u to be $s \cap (\{u\} \times M)$, considered as a subset of M. We say $u \in B$ is a regular point for s if there is a neighbourhood U of u in B and a homeomorphism $h: U \times M \longrightarrow U \times M$ such that $\pi h = \pi$ on $U \times M$ and such that $h(s \cap (U \times M)) = U \times s_u$. We define the catastrophe set of s to be the set of non-regular points in B. It is in general only the catastrophe set of the process which one actually observes in nature.

Of course this definition of a process is much too general to be of use, so we must make some restrictions. We suppose there is given a smooth vector field X on $B \times M$ which is everywhere tangent to the fibres. If $u \in B$ we define X_u to be $X|(\{u\} \times M)$ considered as a vector field on M. We now require a process s to fulfill the condition that s_u is always an attractor of X_u (or is empty). Usually there are additional restrictions of various types (but there is no standard single set of additional restrictions): for example we may require s_u to be always the attractor with the "largest" basin, in some sense of "largest". Or we may require that on an open dense set in B the set s_u always be a stable attractor; in addition we define for u' near u which attractor of $X_{u'}$ "corresponds", i.e. is near, to a given stable attractor of X_u and we also require that if s_u is a stable attractor, then for u' near u the attractor $s_{u'}$ of $X_{u'}$ should "correspond" under this definition to the attractor s_u of X_u.

Now even this situation is very difficult to study, since not enough is known about such vector fields; for example, little is known about when such vector fields have attractors or stable attractors. However the situation becomes much simpler if we make a further restriction: we suppose we are given a smooth function $V: B \times M \rightarrow \mathbb{R}$; for each $u \in B$ we set $V_u = V|(\{u\} \times M)$ considered as a smooth function on M, and we assume that for each u the vector field X_u on M is the gradient field of $-V_u$. In other words we consider V as a family of potential functions on M; the attractors of X_u are then just the local minima of V_u. Here again we require a process s to fulfill the condition that s_u is always a local minimum of V_u (or is empty). The mathematical models in Thom's theory which arise from this situation are called gradient models; the catastrophes which occur (if $\dim B \leq 4$) for the gradient models are called elementary catastrophes. (For Thom's list (theorem 5.6) certain additional physically relevant restrictions are made on the elementary catastrophes for reasons explained at the end of this appendix).

In the case of gradient models too one usually puts additional restrictions on the processes which may arise. There are several conventions which are used in the applications to define such restrictions. One of these is the Maxwell convention, which requires that s_u always be a lowest local minimum of V_u. Another is the convention of perfect delay, which one might describe as follows: To apply this convention, one supposes given, for some manifold B', a diffeomorphism between B and $B' \times \mathbb{R}$ (the factor $\mathbb{R}$ being interpreted as the time direction in B). For convenience, we suppose in the following that $B = B' \times \mathbb{R}$. If s is a process on B, then for each $b \in B'$ we may define a process s(b) on $\mathbb{R}$, with the same state space M, by setting $s(b) = S \cap (\{b\} \times \mathbb{R} \times M)$, considered as a subset of $\mathbb{R}\times M$. The convention of perfect delay requires the following: For each $b \in B'$ and for each $t_0 \in \mathbb{R}$, if there is a process $s' \subset \mathbb{R}\times M$ such that $s'_t = (s(b))_t$ for all $t < t_0$ and such that s' is regular at t_0, then $s(b)$ is regular at t_0. (It is important to caution the reader that this is only one possible explanation of this convention. Thom often does not give detailed definitions of the conventions or of other concepts which are involved in his theory, so that formal descriptions which differ slightly from those we give here can nonetheless be consistent with Thom's writings and may in fact be more useful for some applications).

These conventions may be interpreted as rules which specify when a process s is allowed to "jump" from one minimum of V_u to another as u moves about B; in other words, they are rules for specifying the catastrophe set defined by the family V. The Maxwell convention says that s must "jump" as soon as another minimum becomes lower than the one in which s was originally. The convention of perfect delay says that s must remain in the same minimum as long as possible, that is, until that minimum becomes a point of inflection and disappears. One usually requires that when this happens, the process s jumps to a neighbouring minimum, the one into whose basin the original minimum disappears. There are other conventions, too; for example, there is a convention of imperfect delay which allows s to jump from one minimum to another when the threshold between the two minima becomes low enough; the idea is that "thermodynamic noise" will push s out of the basin of the original minimum when this basin becomes too small.

Of course these conventions do not all associate the same catastrophe set to a family V of potential functions, nor need any one of these conventions define a uniquely determined process associated to V (although in most cases the Maxwell convention yields a uniquely determined catastrophe set; the other conventions need not even do this).

This description of the Thom theory has of necessity been rather brief. For a good understanding of the theory one would need to have many more details, and it would above all be necessary to discuss many examples of the applications; this is beyond the scope of this appendix. However, the description we have given should suffice to make clear what mathematical questions are posed by this theory.

Clearly one of the major problems which suggest themselves in the gradient model case is the problem of classifying the families of potential functions on $B \times M$. Now this is a global classification problem; as a first step in solving it one may consider the easier local problem, i.e. the problem of classifying germs of families of potential functions. In fact, reducing to the local question is not as great a restriction as it may seem; the models given by the Thom theory are only intended to be local models for natural processes anyway; a global description is obtained by piecing together a large number of such local descriptions.

Now a germ of such a family V of potential functions is of course nothing more than an unfolding $f \in \mathcal{E}(n+r)$, where $n = \dim M$ and $r = \dim B$. The importance for Thom's theory of the fibre-space structure of $B \times M$ is reflected in the theory of unfoldings by the importance there of the way $\mathbb{R}^{n+r}$ is fibred as $\mathbb{R}^n \times \mathbb{R}^r$; recall that morphisms and equivalences of unfoldings were all required to respect this fibration.

So the first question it is natural to consider is: given a germ of a potential function on the fibre above a point $u \in B$, what can the nearby potential functions look like? Or in the terminology of this paper: given a germ $\eta \in m(n)$, what r-dimensional unfoldings can η have? This question was answered, at least for finitely determined η, in Chapter 3. There we gave a characterization of universal unfoldings and in fact gave a method for constructing a universal unfolding of any finitely-determined germ η; any other unfolding of η is induced from a universal unfolding by mappings which respect the fibration of $\mathbb{R}^{n+r}$.

In general, of course, a given germ will have infinitely many non-isomorphic unfoldings, and if the germ is not finitely determined one does not even know what unfoldings it can have. So it is natural to ask whether all of these unfoldings are actually relevant to Thom's theory; can they all actually occur in describing real events observable in nature?

The answer, according to Thom, is no. In descriptions of natural phenomena by ("stable") gradient models only seven essentially different unfoldings can be involved; the catastrophes they define locally are the seven elementary catastrophes, and according to Thom they describe a large class of natural phenomena, including phenomena involving the propagation of wavefronts and many important biological phenomena.

In our Theorem 5.6 we gave one formal mathematical interpretation of Thom's famous list of the seven elementary catastrophes. The motivation for the hypotheses and the formulation of this theorem should now be clear. The hypothesis that the unfolding f be stable corresponds in Thom's theory to the requirement that f be the local description of an event observed in nature. An event is "observable in nature" only if it can (at least in principle) be produced repeatedly in numerous experiments; but since the same set of initial states and the same set of environmental conditions for an experiment can never be reproduced exactly, this means that the essential structure of a natural process must be invariant under small perturbations. In other words the family of potential functions which generates the process must be stable.

Recall that Thom usually takes B to be physical space-time. Of course in some applications it is enough to assume that B is a submanifold of space-time, for example when one is studying an event extended in space but at a single moment of time, or when one is studying the temporal evolution of a process at a single point in space. In any event, one usually is in a situation where B is at most four-dimensional. This explains the hypothesis of theorem 5.6 that the unfolding dimension be $\leqslant 4$. (Incidentally, for applications in which B has dimension greater than four it might be interesting to have an analogue of Thom's list which would cover stable unfoldings whose unfolding dimension are small but not necessarily $\leqslant 4$. Such a generalization of Thom's list can easily be obtained using the full statement of Mather's classification of germs of right-codimension $\leqslant 5$ [10, Ch.II] and using the work of Siersma [12]).

Finally it is clear that there is no point in including in the classification unfoldings which do not generate observable events, even though they may satisfy the first two hypotheses. Recall that it is only the catastrophe set of a process which is actually observed. In particular this means that an unfolding which allows no non-empty processes is uninteresting from the point of view of Thom theory; but non-empty processes can occur locally only if there are nearby points where the potential functions V_u have local minima. This is why in Theorem 5.6 we require f to have local minima near 0, and this is why it is important to know that f reduces with index 0 to an unfolding in the list, since otherwise f cannot have local minima near 0. Incidentally these considerations also explain why the case of a simple minimum (or more generally of a simple singularity) is excluded from the list and mentioned separately, although an unfolding with a simple minimum does have minima near 0. Recall that if $f \in \mathcal{E}(n+r)$ unfolds $\eta \in m(n)$ and if f has a simple minimum at 0, then f is orientedly equivalent to the r-dimensional constant unfolding of $Q_o \in m(n)$, where $Q_o(x) = x_1^2+\ldots+x_n^2$. Hence in each fibre near 0 there is exactly one minimum; and no matter which of the standard conventions we use, a non-empty process must intersect each fibre in this unique minimum. So locally every process is regular, and there can be no catastrophe points nearby. Therefore this case is uninteresting.

Note that in Theorem 5.6 there are no restrictions on the size of n (which corresponds to the dimension of the state space). This has the important consequence that even if a process depends on a very large number of physical parameters (as is often the case in the applications), as long as it is described by a gradient model its description will involve one of the seven elementary catastrophes; in particular one can give a relatively simple mathematical description of such apparently complicated processes even if one does not know what the relevant physical parameters are or what the physical mechanism of the process is. And the number of parameters which are involved in the physical mechanism plays no role in the description.

REFERENCES

[1] J.M. Boardman, Singularities of Differentiable Maps, prepublication version (1965) (published in an emasculated form in IHES Publ. Math. No. 33 (1967), pp. 21-57.)

[2] D. Gromoll and W. Meyer, On differentiable functions with isolated critical points, Topology 8 (1969), 361-369

[3] B. Malgrange, Ideals of Differentiable Functions, Oxford University Press, London, 1966

[4] J. Mather, Stability of C^∞ mappings: I. The division theorem, Ann. of Math. (2) 87 (1968), 89-104

[5] J. Mather, Stability of C^∞ mappings: II. Infinitesimal stability implies stability, Ann. of Math. (2) 89 (1969), 254-291

[6] J. Mather, Stability of C^∞ mappings: III. Finitely determined map-germs, Publ. Math. IHES 35 (1968), 127-156

[7] J. Mather, Stability of C^∞ mappings: IV. Classification of stable germs by $\mathbb{R}$-algebras, Publ. Math. IHES 37 (1969), 223-248

[8] J. Mather, Stability of C^∞ mappings: V. Transversality, Advances in Math. 4 (1970), 301-336

[9] J. Mather, Stability of C^∞ mappings: VI. The nice dimensions, in [18], pp. 207-253

[10] J. Mather, unpublished notes on right equivalence.

[11] J. Milnor, Morse Theory, Ann. of Math. Studies 51, Princeton University Press, Princeton, 1963

[12] D. Siersma, The singularities of C^∞-functions of right-codimension smaller or equal than eight, Indag.Math. 35 (1973), pp. 31-37

[13] R. Thom, Un lemme sur les applications différentiables, Bol.Soc.Mat.Mex (2) 1 (1956), 59-71

[14] R. Thom, Topological models in biology, Topology 8 (1969), 313-335

[15] R. Thom, Stabilité Structurelle et Morphogénèse, W.A.Benjamin, Inc., Reading, Massachusetts, 1972

[16] R. Thom, Modèles Mathématiques de la Morphogénèse, Ch. 3: "Théorie du déploiement universel", IHES (1971), (unpublished)

[17] J.-C. Tougeron, Idéaux de fonctions différentiables I, Ann. Inst. Fourier 18 (1968), 177-240

[18] C.T.C. Wall, ed., Proceedings of Liverpool Singularities Symposium I, Springer Lecture Notes in Mathematics 192, Springer-Verlag, Berlin, 1971